Exploring Careers in Cybersecurity and Digital Forensics

Exploring Careers in Cybersecurity and Digital Forensics

Lucy K. Tsado
Robert Osgood

ROWMAN & LITTLEFIELD
Lanham • Boulder • New York • London

Published by Rowman & Littlefield
An imprint of The Rowman & Littlefield Publishing Group, Inc.
4501 Forbes Boulevard, Suite 200, Lanham, Maryland 20706
www.rowman.com

6 Tinworth Street, London, SE11 5AL, United Kingdom

British Library Cataloguing in Publication Information Available

Library of Congress Cataloging-in-Publication Data

Names: Tsado, Lucy K., 1970– author. | Osgood, Robert, 1958– author.
Title: Exploring careers in cybersecurity and digital forensics / Lucy K. Tsado, Robert Osgood.
Description: Lanham, Maryland : Rowman & Littlefield, [2022] | Includes bibliographical references and index. | Summary: "Exploring Careers in Cybersecurity and Digital Forensics serves as a career guide, providing information about education, certifications, and tools to help those making career decisions within the cybersecurity field"—Provided by publisher.
Identifiers: LCCN 2021023421 | ISBN 9781538140611 (cloth) | ISBN 9781538140628 (epub)
Subjects: LCSH: Computer security—Vocational guidance. | Digital forensic science—Vocational guidance.
Classification: LCC QA76.9.A25 T763 2022 | DDC 005.8—dc23
LC record available at https://lccn.loc.gov/2021023421

This book is dedicated to the students who are seeking career opportunities that will enrich their lives. Your success is our success, and we hope this book can help get you to where you want to be.

Contents

Preface

We wrote *Exploring Careers in Cybersecurity and Digital Forensics* for three reasons: to help students and school administrators understand the opportunity that exists in the cybersecurity and digital forensics field; to provide guidance for students and professionals looking for alternatives through degrees, be it criminal justice, business, computer science, or engineering; and to help by closing the cybersecurity skills gap through student recruiting and retention in the field. We understand that the cybersecurity and digital forensics pathways are an alternative for many who wouldn't normally resonate with those programs but do want to be involved. *Exploring Careers in Cybersecurity and Digital Forensics* was written as a guide for these students. Moreover, as the use of technology continues to grow, taking the proper precautions to monitor digital criminal activity is now pertinent. Secondly, with the rise in cyberattacks and cybercrimes, there is an increased demand necessary to prevent, detect, and respond to breaches as well as gather and preserve evidence.

The skills gap in cybersecurity is a wonderful opportunity for anyone who is passionate about and appreciative of the issue of criminal behavior in cyberspace. Cybersecurity and digital forensics are perceived as technology-oriented fields, but they also involve the collaboration of other disciplines and thus many different fields of study. It is a general myth that the cybersecurity field is only suitable for those in the fields of information technology and computer science. This is not entirely true because professionals are calling for other technical and nontechnical academic fields to get involved in the cybersecurity workforce. As we teach criminal justice and engineering students courses related to cyber defense, many have asked how they can enter the field of cybersecurity and digital forensics. This book was designed to be that guide and roadmap. Since several students struggle with how to successfully enter the field and require guidance in making these decisions, *Exploring Careers in Cybersecurity and Digital Forensics* was designed to offer career guidance to students and counselors in high schools and colleges. Even though this book is written from a criminal justice perspective, it can be used by anyone as long as the interest and passion are there.

The cybersecurity skills gap has developed and is worsening due to the slow response to the issue at hand. This book addresses how to close the skills gap by creating a pipeline of talent through education and training. *Exploring Careers in Cybersecurity and Digital Forensics* therefore addresses the skills gap from the unique perspective of attracting nontraditional students that would typically not be involved within the cybersecurity, cyber defense, or digital forensics field. This book targets students and those who advise them, from human behavioral disciplines like criminal justice and psychology to legal disciplines such as law. Additionally, nontraditional segments of business and engineering may also not be aware that their skills are relevant and necessary in the cybersecurity/digital forensic field since they are not usually drawn to the field. *Exploring Careers in Cybersecurity and Digital Forensics* will serve as a career handbook for such students and advisors, providing information about education, certifications, and tools to guide them in making career decisions within the field.

Acknowledgments

We would like to thank Rowman & Littlefield, our publisher, for the support for and belief in this project. We would also like to thank Kathryn Knigge for seeing this as an opportunity and helping us through those first few months when we didn't even think this was a possibility. Finally, many thanks to Becca Beurer and her team members for their patience and guidance and for their aid with the last stretch and completion of the project, helping us over the finish line. We couldn't have done this without all of you. Thank you!

CHAPTER 1

What Is Cybersecurity?

Defining cybersecurity is difficult. Consequently, finding one definition of cybersecurity is impossible. Most importantly, cybersecurity means different things to different people. For example, according to the Department of Homeland Security (DHS), as of 2016 cybersecurity was "the activity or process, ability or capability, or state whereby information and communications systems and the information contained therein are protected from and/or defended against damage, unauthorized use or modification, or exploitation." They have since updated that definition because it was inadequate to explain the full scope of cybersecurity. The updated definition, which they described as an extended definition, describes cybersecurity as

> strategy, policy, and standards regarding the security of and operations in cyberspace, and encompass[ing] the full range of threat reduction, vulnerability reduction, deterrence, international engagement, incident response, resiliency, and recovery policies and activities, including computer network operations, information assurance, law enforcement, diplomacy, military, and intelligence missions as they relate to the security and stability of the global information and communications infrastructure.[1]

The initial definition, specific to the Department of Homeland Security and its operations, did not include information about responses to digital threats or recovery of compromised assets. One of DHS's primary

1

tasks is keeping critical infrastructure (CI) safe. However, the updated extended definition captures this.

Two leading cybersecurity firms, Kaspersky and the Palo Alto Networks, have similar definitions for cybersecurity. Kaspersky defines it as "the practice of defending computers, servers, mobile devices, electronic systems, networks, and data from malicious attacks,"[2] while Palo Alto Networks states that cybersecurity "refers to the preventative techniques used to protect the integrity of networks, programs and data from attack, damage, or unauthorized access."[3]

Perhaps Norton's definition, which hints on recovering data from cyberattacks specifically, is more definitive of cybersecurity even though it is still incomplete. Norton states that cybersecurity "is the state or process of protecting and recovering networks, devices, and programs from any type of cyberattack."[4]

It is evident that focusing on only one definition of cybersecurity will not serve the purpose of dealing with what cybersecurity is. We do not attempt to define cybersecurity due to the challenges involved. We do, however, state that cybersecurity involves many encompassing parts. It involves protecting and defending data and systems from unauthorized access, exploitation, damage, or use; recovering data and systems where an unauthorized access, exploitation, damage, or use has occurred; and restoring data and systems *after* an unauthorized access, exploitation, damage, or use has occurred. Cybersecurity also involves risk analysis, incidence response, and business continuity planning. Basically, information security, a subset of cybersecurity, has three fundamental foundations: confidentiality, integrity, and availability, known as the CIA. Any definition that ignores these three foundational concepts is incomplete as to how to approach or deal with cybersecurity.

Therefore, any working definition of cybersecurity is a function of what it means to an organization or an industry. Consequently, this book will address cybersecurity from a criminal justice standpoint in responding to criminal behavior that involves the use of information and communication technology (ICT) and/or the use of cyberspace in the commitment of a crime. Kremling and Sharp Parker compiled definitions associated with various laws and the Department of Homeland Security and concluded that defining cybersecurity in the realm of criminal justice will involve the jurisdiction in which the criminals are operating, the tactics of a perpetrator or perpetrators, and the prevention techniques needed to prevent the activities of cyber criminals from materializing.[5] Therefore, these will be the focus of the discussions in this book.

Cybersecurity did not always exist as an academic field. The proliferation of ICT has facilitated cybercrime, requiring the defense of cyber resources to include critical infrastructure. As cyberattacks and cybercrimes became more prevalent, so did the need to protect against these threats. It is important to understand that cybersecurity is still evolving as an academic field.

Cybersecurity and the Criminal Justice Connection

The connection between cybersecurity and criminal justice has been established, but for the purpose of clarity (or to avoid ambiguity), it needs to be discussed. We use the terms "cybersecurity" and "digital forensics" in this book interchangeably to refer to career choices students can make. While cybersecurity is a wide career field, digital forensics is specific. We discussed earlier that an attempt to define cybersecurity is difficult; however, we can set a threshold by determining that cybersecurity involves *preventing* criminal behavior in cyberspace or by the use of ICT, *responding* to behavior that is criminal that involves cyberspace or the use of ICT, and *recovering* from an attack involving cyber-criminal activity. While all three relate to criminal justice, the one most related to criminal justice is response.[6] By nature, law enforcement is called upon to respond to criminal behavior in the physical realm, and so it is in cyberspace, thus the connection to digital forensics.

There is a need to respond to criminal activity, whether it is in the physical realm or in cyberspace. When cybercrimes occur, there is a need to respond, hence the connection to criminal justice. What makes this confusing, however, is the fact that many cybercrimes occur in private companies. Private companies are sometimes unwilling to report such activities to avoid losing credibility or clientele. However, when cybercrimes involve government entities, there is a required need to respond. Many times, this response falls on the Federal Bureau of Investigation (FBI) and other agencies like the U.S. Secret Service (USSS), DHS, and other such agencies. In some jurisdictions, local law enforcement agencies are charged with responding. Due to the ubiquitous nature of cybercrimes, jurisdiction can be an issue. However, once a victim is in a U.S. jurisdiction, the United States responds.

Ultimately, response to cybercrimes depends on what type of cybercrime has occurred and the response needed to address it. It is therefore important to examine the different types of cybercrimes and how they are responded to. David Wall categorized cybercrimes into four main groups: cyber trespass, cyber deception/theft, cyberpornography and obscenity, and cyberviolence. Wall states that cyber trespass involves unauthorized crossing of already established boundaries in cyberspace. Cyber deception/theft, on the other hand, involves using cyberspace to steal or cause harm. Cyberporn and obscenity involve cases where sexually expressive materials are traded in cyberspace, while cyberviolence includes violence and the subsequent impact that violence has on people using cyberspace. It could be carried out by an individual or a social or political group against another.[7] After Wall's typologies as determined in 2001, other types of cybercrimes have evolved that need to be addressed.

These categorizations are important because they allow law enforcement to devise strategies and tactics to respond to the various types of cybercrimes. There is therefore a clear connection between cybersecurity and criminal justice. Going by Kremling and Sharp Parker's description of the definition of cybersecurity, the connection between cybersecurity and criminal justice can be expanded from response as is done by law enforcement to include defense, protection, and recovery in other federal government agencies and in the private sector. These will be addressed in chapter 2.

The Evolution of Digital Forensics

Humans have been using devices to help them calculate and organize things for thousands of years, but the first devices that we can confidently call computers appeared during World War II in the 1940s. These were massive machines that filled rooms. Today the Apple Watch has more computational power than these machines.

In the 1970s, we read Donn Parker's book *Crime by Computer*, and we were fascinated by the ways criminals could use and exploit computers for the purpose of committing crimes. But the true beginning of the age of computer forensics, now called digital forensics, was in the early 1980s. So what happened in the early 1980s that prompted the rise of computer forensics? The answer is the creation of the IBM PC and the Apple Macintosh a few years later. These computers were much easier to use than their mainframe or mini-computer ancestors and were affordable, so the average person, small business, or criminal could own one.

At first, computers were being seized by law enforcement in more traditional white-collar crime cases where criminals were using computers to perpetrate their fraud: create and store documents, track finances, and so on. By the late 1980s, we were seeing computers used in crime in different ways.

It was the summer of 1989. It was very hot, and a team of FBI agents were executing a search warrant on an apartment in Southern Towers just off of Seminary Road and Interstate 395 in Alexandria, Virginia. The apartment was the home of Daniel Thomas DePew. DePew was arrested along with Dean Lambey and charged with conspiracy to kidnap in violation of 18 USC S 1201(c) and conspiracy to sexually exploit a child in violation of 18 USC S 371.[8]

The team of FBI agents entered the apartment and executed the warrant. DePew's Macintosh computer was seized along with other items of evidentiary value. It was DePew on his Mac chatting online through a bulletin board with undercover police officers that alerted the FBI that he and Lambey planned to kidnap a child outside of a Northern Virginia school, film the child being sexually abused, subsequently murder the child (otherwise known as making a snuff film), and then sell the film on the underground child pornography market. In 1989, DePew was using the internet to orchestrate this heinous act, and this was four years before the rollout of the World Wide Web (although, coincidentally, 1989 was the same year that Tim Berners-Lee invented the World Wide Web).[9] Now computers were being used by criminals, in this case heinously violent criminals, to communicate via bulletin boards.

The DePew/Lambey case was also the first case in the United States where a digital intercept order was obtained tapping the computers of the subjects communicating on bulletin boards. Bulletin boards are passé today, but the internet through Tor and the dark web is a prolific conduit for criminal communication and commerce.

The 1990s saw a new type of digital crime. The cellular telephone, aka the flip phone, manufactured by companies such as Motorola, made a significant impact on society. Now criminals were using cell phones to communicate and with the advent of cloned phones were communicating with virtual anonymity. A phone cloner would stand near a high-traffic area (e.g., the George Washington Bridge in New York City) with a hand scanner and collect the phone numbers and electronic serial numbers (ESN) of the phones passing by in cars. The phone cloner would then use this information to modify cellular phones to use these stolen numbers. Now we have computers being used to steal telecommunication services. These stolen cell services are purchased by drug dealers to conduct their illegal business. Once the legitimate owner of a phone number received her bill and saw all the additional calls and charges, that phone number would be cancelled, and the drug dealer would need to buy a new cloned phone number.

With 3G being rolled out in the United States in 2002, cell phones are now cheap enough for the masses and can handle data as well. Essentially, everyone is carrying a computer in their bag or pocket. These mobile devices make phone calls (how old is this), send/receive text messages and emails, stream video, and are able to run an unlimited number of applications or apps (Twitter, Snapchat, and WhatsApp, to name a few). This use creates massive amounts of data that is stored on the phone, as well as somewhere called the cloud (more later on this). Techniques were developed to forensically extract data from mobile devices. Such tools as Cellebrite, XRY, Paraben, MOBILEdit, and Oxygen are used to extract evidence from mobile devices.

In January 2002, thirteen-year-old Alicia Kozakiewicz was lured out of her home in a suburb of Pittsburgh, Pennsylvania, by Scott Tyree, thirty-eight, and taken to Herndon, Virginia, where the FBI found her a few days later chained to a bed.[10] Tyree was arrested at his place of employment. Kozakiewicz struck up an online relationship with Tyree, and Tyree convinced her to leave her family and go with him. Kozakiewicz walked out of her home, got into Tyree's car, and left. The young girl was lucky. The digital evidence found on her home computer led the FBI to Tyree's townhouse in Herndon. Although in its infancy, social media, the ability of people to interact with each other through the internet (a generally positive experience), was being used by criminals to exploit children.

Back to this "cloud thing," cloud computing, essentially using someone else's computer remotely, has been around since Moses was a junior programmer at IBM, but modern cloud storage capability offered to anyone has really evolved during the past ten years. In 2011, for example, Apple launched iCloud so anyone with an Apple device (Mac, iPhone, iPad) now has access to remote storage seamlessly integrated with their device.[11] Amazon Web Services (AWS), Microsoft Azure, and Google Cloud are just a few, albeit quite large, players in the cloud industry today. Investigators now must deal with not only devices that criminals have physical access to but the cloud as well.

What about the world of IoT, or the Internet of Things? Our toasters and refrigerators are getting smart. Alexa and Siri monitor you to respond to your wishes, and anyone can buy a drone with a camera built into it. What this all leads up to is that just about everything today that has a chip in it may be used for or in conjunction with criminal activity. History has proved this time and time again.

Last but not least, there is weaponized code, aka malware or the virus/worm. The first publicly known weaponized code attack occurred in 1988, when Robert Morris Jr. unleashed his out-of-control code, crippling thousands of computers.[12] We all know that weaponized code attacks are as prolific today as sand on a beach. Investigators using legal tools such as the Computer Fraud and Abuse Act of 1986 must deal with the transnational aspect of this type of crime, the technical complexity, and the anonymity of the internet, as well as public apathy to a certain degree since we are all getting used to these intrusions, with ransomware attacks targeting not only individuals but also hospitals, businesses, and government agencies.

Today, digital forensics is an integral part of law enforcement. Virtually every investigation has some digital nexus, and with this digital nexus comes the need for digital forensics examiners to collect, process, analyze, and potentially testify to evidence in a court of law. Career opportunities are available in both the sworn officer and forensic examiner roles.

CHAPTER 2

The Cybersecurity Skills Gap

AN OPPORTUNITY FOR
CRIMINAL JUSTICE STUDENTS

There is currently a critical skills gap in the cybersecurity field. Many have been calling attention to this skills gap since it became obvious that there was a need to identify/defend/respond to cyberattacks. In 2010, research strategists at the Center for Strategic and International Studies in a report titled "A Human Capital Crisis in Cybersecurity: Technical Proficiency Matters" stated that "the cyber threat to the United States affects all aspects of society, business, and government, but there is neither a broad cadre of cyber experts nor an established cyber career field to build upon, particularly within the Federal Government."[1] While the situation has improved at the time of this writing, as per the structuring of the cybersecurity academic field, the dire need for professionals still exists today. What has changed is the partnership between the National Institute of Standards and Technology (NIST) and DHS, which created National Initiative for Cybersecurity Education (NICE) to structure the academic field, enhance the quality of professionals entering the workforce, and generally streamline the cybersecurity education and workforce arena.

Despite this move, the continuous proliferation of various technological innovations in both government and private sectors, as well as the ensuing opportunity created for criminal activity in ICTs and cyberspace, has amplified the demand for cybersecurity professionals. Therefore, the need to fill existing cybersecurity jobs—and the fact that there is 0 percent unemployment in the field as well[2]—makes it a great career move for anyone and especially criminal justice and other majors, if they have the right skill set.

Criminal Justice Students and the Infusion of Cyber Forensic Skills

The belief that cybersecurity is a career option for only those in technical fields is mistaken. Gone are the days when cybersecurity was only meant for people with technical skills alone. Practically anyone with passion, interest, and means can join the field. As earlier discussed, the fact that response to criminal activity in cyberspace is needed creates an opportunity for criminal justice students within the cybersecurity profession. Therefore, the opportunity exists for anyone.

There are several reasons why the cybersecurity profession in general is a great career choice for anyone. It is a growing and broad academic field, and there are facets that those without a technical background can still join.[3] Particularly for the criminal justice field, cybercrimes are increasing daily, and there is a need to respond to them. This means that there will continue to be a need for digital forensic examiners, which is the main cybersecurity career choice associated with criminal justice. There are other career paths that provide opportunities especially within the private sector, like information technology auditor and cybersecurity consultant, to mention a few. Consequently, now is the time for criminal justice and other nontechnical students to take advantage of this career opportunity.

As stated earlier, there is currently a 0 percent unemployment rate in the cybersecurity field. This means for the next few years, the need exists for cybersecurity professionals as there are many government and private sector jobs going unfilled. According to the Bureau of Labor Statistics (BLS), there is a growing need for information system analysts, with an anticipated growth rate of 32 percent from 2018 to 2028.[4] The need for digital forensic examiners is not expressly determined by the BLS probably because it has not yet been separately recognized as an occupation. However, with the continued growth in the use of ICTs, the use of the Internet of Things (IoT) technology to make life comfortable for people, and the continuous increase in cyberattacks and cybercrimes, it is safe to say there will be a need to respond, hence the need for digital forensic examiners and cybersecurity analysts. (An in-depth discussion on digital forensics will be carried out in chapter 3.)

Secondly, cybersecurity is a fairly new academic field. Unlike many other academic fields, like medicine, engineering, and computer science, that are structured, well planned, and well established, cybersecurity is still evolving. There were arguments in the past to make it a rigorous academic field like medicine and engineering.[5] However, these arguments, though well founded, have been overtaken by events due to the dire need for cybersecurity professionals. The Center for Academic Excellence (CAE) designation is one way that this problem has been addressed. By identifying and designating institutions of higher learning as CAEs, the education of cybersecurity professionals is now fairly structured, and the number of professionals in the field will hopefully increase. It is important to note that not all institutions that teach cybersecurity are CAE designated. Therefore, there are other ways to teach cybersecurity without being a CAE-designated school. The CAE designation is just a way to identify and designate educational institutions that have been evaluated to teach cybersecurity by the DHS. There are other universities and institutions of higher learning that teach cybersecurity. There are also informal avenues to get cybersecurity training and certifications, from companies like CompTIA, (ISC)[2], ISACA, and many such organizations. (These will be discussed in chapter 4).

While many cybersecurity jobs are technical in nature, there are also other areas in cybersecurity that are not. Areas in cybersecurity that involve planning and risk management, for example, don't need curricula steeped in technical knowledge. NIST has mapped out educational programs and degrees for institutions willing to teach cybersecurity. According to the National Information Assurance Education and Training Program (NIETP),[6] there is a technical and a nontechnical knowledge unit. The technical unit involves knowledge components like scripting, programming, networking, cryptography, and operations systems concepts. The nontechnical core involves knowledge units like policy, legal, ethics, and compliance; cyber threats; security program management; security risk analysis; and security planning and management. It is important to note that there are foundational knowledge units like cybersecurity foundations, cybersecurity principles, and IT systems components that are a part of all cybersecurity programs.[7] These foundational knowledge units are intended to provide nontechnical students the needed foundational knowledge upon which they can build their nontechnical core to be successful cybersecurity professionals. But students will also need some technical skills as well.

The dire nature of the need for professionals in the field is making other avenues available for cybersecurity education and training. Formal education through associate, bachelor's, and master's degrees provides students with foundational and specific knowledge cores that prepare students for rewarding careers in cybersecurity, while short-term training programs and certifications provide a means to join the field through practical, firsthand, real-world training. There is currently no way of determining if a formal or informal education is better. What we do know is that both are important for a well-rounded career in cybersecurity. For example, what good is a professional who does not understand the risk threshold of their organization but has all the technical knowledge? Or what good is a professional who does not appreciate the implication of having a business continuity plan as part of the risk assessment of a company? What good is a professional who can code without understanding that current applications need to have built-in security and how this is important for organizational continuity? Or what good is a professional who knows all the risks but does not understand the basic foundation on how to mitigate those risks? Therefore, both formal (education) and informal (training and certifications and other short-term avenues) benefit a robust cybersecurity career. It is our opinion that both formal and informal education are needed to be successful in the cybersecurity field.

Organizations and Law Enforcement Agencies Will Need Cybersecurity Students

Organizations must understand what their risks are and what is needed in terms of tools and people (resources) to mitigate those risks. They should also understand how to go about developing, obtaining, and sustaining these resources. Organizations need to understand that cybersecurity is no longer left to the relics of core operations, but it is an important business function of any organization worth its salt. It should be treated as an important business function of an organization and, depending on the organization's core competence, separate from information technology, just like auditing is from accounting.

Organizations as discussed here could also include law enforcement agencies who have to decide what their response team should look like. As discussed in chapter 1, the FBI and U.S. Secret Service are the federal agencies and the lead investigating authority for cybercrimes in the United States. There are other local and state agencies that also respond to cybercrimes and have the duty to respond due to the increasing number of breaches occurring. Each state has an agency that responds to cybercrimes. Some states have more than one agency. Therefore, with the continuing increase of breaches against government agencies, there will be jobs for criminal justice professionals interested in cybersecurity and digital forensics. In 2019 alone, there were breaches in many state and local governments as seen with the following headlines:[8]

"22 Texas Towns Hit with Ransomware Attack in 'New Front' of Cyberassault"[9]
"Hackers Are Holding Baltimore Hostage: How They Struck and What's Next"[10]
"Hit by Ransomware Attack, Florida City Agrees to Pay Hackers $600,000"[11]
"Mississippi City Operations Disrupted by Ransomware Attack"[12]

We foresee a future where organizations including law enforcement agencies will start going into prevention as a tactic to reduce their exposure to breaches. Breaches like these will create jobs for criminal justice professionals who have risk, threat, and incident planning and business continuity planning knowledge.

What Educators, Advisors, and Career Counselors Need to Know

Educators of cybersecurity professionals have an important role to play in the success of those wishing for a career in cyber forensics. Educators as discussed here are not limited to teachers and professors but include others in decision-making roles about academic programs, as well as those in advisory and counseling roles. Educators' roles in creating cybersecurity professionals are slightly different from other career fields due to the dynamic nature of the cybersecurity field and the critical need in the industry. Educators need to understand that the field is different from other career areas, and therefore the approach to identifying, attracting, training, and sustaining cybersecurity talent for their programs has to be different. In fact, cybersecurity education as an institution is best addressed first by addressing what industries the institution serves. For example, Houston's main industries are oil and gas and the medical field. It would make sense for academic programs in the area that want to educate cybersecurity professionals to identify what the needs of these industries are and create programs that will feed these industries. We are not suggesting limiting the academic programs to serving only the immediate vicinity, but we do realize that it makes sense to have a mechanism in place that will feed local as well as state and national arenas.

The vibrancy of a cybersecurity academic program comes from industry's recognition of that program as a source for qualified cybersecurity professionals. This means that academic institutions will need to create a cybersecurity ecosystem that has a plan to involve all members of its community in cybersecurity education. Therefore ideally, cybersecurity education should be introduced from the elementary school level and continued in the middle and high school so that there is a continuous flow of students into higher education programs. Institutions of higher education should be involved in that process. While these efforts are being made by NICE and its partners, the impact is yet to be felt because of the slow pace. Educational institutions should also have relationships with partners in the industry who will be willing to hire from their schools, provide internships and apprenticeships, as well as provide experiential learning opportunities to students for hands-on learning experiences. This will enable an educational institution to create and maintain a vibrant cybersecurity ecosystem.

Educators also need to ensure that students are exposed to cybersecurity certifications as this has proved to be an important factor in the hiring decision of professionals. (ISC)[2] is a nonprofit organization that is involved in cybersecurity training; it carried out a survey of cybersecurity professionals in the field and found in 2018 that 43 percent of respondents stated that having certifications was important in their hiring decision.[13] Consequently, it is important that students are introduced to organizations like Coursera, Cybrary, CompTia, (ISC)[2], ISACA, and Infosec and learn where they can obtain hands-on experience by way of certifications and training. Hence, while some cybersecurity formal educational programs (bachelor's and master's) may be adequate for getting hired, it is not enough to sustain a cybersecurity career. The need for certifications is important because of the dynamic nature of the field. Technological changes are evident in that twenty years ago there was no widespread use of IoT. IoT has become a core competence/supporting technology of many technological companies; therefore the need to secure systems due to these interconnectivities is paramount, as well as the need to defend, respond to, and recover from breaches.

How Can a Student Attain a Successful Cybersecurity Career?

Obviously, there are currently opportunities for just about anyone interested in the cybersecurity field, not just criminal justice majors. While this opportunity exists, it comes at a price. A career move toward cybersecurity will require a passionate participant, the required resources, and some guidance, the most important of these three being the passion. A passionate participant is required for any career. Anyone who is interested in entering the cybersecurity field must be passionate because it is not an easy feat. Educators can ignite this passion by creating a mechanism that targets and identifies students early at the middle and high school level. At this level, it is easier to identify cybersecurity talent using motivating projects like hackathons and gaming competitions. These activities and projects not only identify talent but also enable young minds to be critical thinkers and problem solvers.

The cybersecurity field is currently looking for students who are not only passionate but also are problem solvers and critical thinkers, those who think outside the box. Many challenges in cybersecurity include the ability to think outside the box and the ability to understand core business functions and how these functions are affected by security. Critical thinkers and problem solvers have the ability to understand these business functions and anticipate where breaches may occur, as well as how to respond to breaches when they do occur. Many criminal justice responsibilities also include these qualities, and therefore criminal justice students will make good cybersecurity professionals especially when they are educated and are adversarial in their thinking.[14] They must be able to think outside the box and problem solve. Therefore, early identification is key to finding those who will be passionate about cybersecurity.

The second ingredient in a successful cybersecurity career move is the resources. There are currently various avenues for resources to get a cybersecurity degree through scholarships like CyberCorps through the DHS.[15] However, these scholarships are limited and mostly only available to students in established cybersecurity programs like CAE-designated schools and veterans. What is also lacking is funding for certifications so that qualified students can attain the certifications that will allow them not only to get their foot in the door but also to differentiate themselves as skilled professionals starting careers in cybersecurity. Certifications are expensive; however, students can take advantage of other avenues like Cybrary and other such free online programs that will give them hands-on experience.

Finally, students need to be guided in making career decisions in the cybersecurity field. Cybersecurity by its current nature is a very wide field, and there are many facets. Students need to be guided as to what aspects of cybersecurity best suit their personalities. Again, identification at an early level of education will help. However, even at the university or higher education level, there needs to be a career center that helps with counseling and guiding students as well as helps them get internships, apprenticeships, and other experiences that will help them decide if a cybersecurity career is what they want. In fact, this is one of the requirements of getting a CAE designation and will be a great move for educators to make. Such career centers are great because they not only guide students through education and career choices, but they also connect students to jobs, internships, apprenticeships, training, certification programs, and the cybersecurity community.

Unfortunately, many seniors in high school do not know what career they want to pursue at the time they are graduating high school. Many more are not aware of what they want to do in their sophomore year of college. This is generally not bad as long as they have an idea by the time they graduate from college and they have accumulated the proper body of knowledge. Involving career centers in their pursuit of what career choices they need to make is important. They need to become involved in career center activities as early as their freshman year in college so that they can begin making career choices and "build their brand," a term Terhune and Hays used to describe how students can build their personal brand by identifying who they want to be and the adjectives they want used to describe them. "Building your brand" also includes being deliberate about career choices, making decisions toward the personal brand every day, and protecting the brand by making sure that the actions of students match the words they have used to describe themselves. They state that every student should strive to be honest and have integrity, which will in turn make them credible. This, they assert, will help students build their personal brand.[16]

CHAPTER 3

It's All about Skills

Digital Forensics Swim Lanes

Digital forensics is not a discipline but a series of nested disciplines with both similarities and differences. These swim lanes coincide with the sources of digital evidence.

DIGITAL MEDIA FORENSICS— THIS LANE HAS BEEN AROUND THE LONGEST

Computers are everywhere. Although desktop computers are going the way of the dodo bird, they still exist with laptops, notebooks, and other similar variations. These devices contain hard drives. These hard drives are in the form of a spinning magnetic disk (also being phased out), SSD or solid-state drives, or solid-state chipsets.

The forensic process involves imaging the target media in a write-protected state (although this is not always feasible), ingesting the image into one of several popular forensic tools (Autopsy, Axiom, Blacklight, Encase, FTK, Nuix, Truxton, etc.[1]), and analyzing the media for its evidentiary value.

The imaging is a bit-by-bit forensic copy of the media that can be verified via hash analysis and can be performed with software or hardware. On the software side, FTK Imager is probably the most widely used tool,

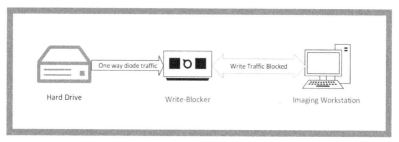

One way diode traffic

Write Traffic Blocked

Hard Drive

Write-Blocker

Imaging Workstation

Figure 3.1.   Write-blocker. *Figure created by authors*

but Linux dd is always a viable alternative. Write-blockers do need to be used to protect the forensic integrity of the target media.

On the hardware side, there are several options. CRU/Wiebetech has the Ditto,[2] Logicube has the Talon or the Falcon,[3] and Open Text/Tableau has the TD2U and TX1.[4] Write-blocking is incorporated into this technology. The hardware-based imaging solutions are generally much faster than software imaging, but they are considerably more expensive. Please remember that time is money, so in a production environment, hardware-based imaging is generally the way to go.

Imaging is usually accomplished with the drive connected to a write-blocker or imaging device and by then energizing the drive in read-only mode (figure 3.1). There are times, however, when you are dealing with production servers that cannot be taken offline, where live imaging or logical imaging may be your only answer. Live imaging is the taking of a bit-by-bit copy of a system that is in production (read/write) mode, meaning that the image to take will be unique and will not hash to the original. The same is true of a logical image where the file system is copied into a forensic container. Both are options when dealing with encrypted file systems.

There is a plethora of operating systems that you will encounter, but the gorillas in the room are Windows (Workstation and Server), Apple's Mac OS, and Linux with Windows holding the majority of the market. By the way, Linux is an excellent forensics platform.

NETWORK FORENSICS—ENCRYPTION IS MAKING NETWORK FORENSICS MORE DIFFICULT

Intercepting network traffic can provide crucial evidence in your case. The legalities of intercepting network traffic are not covered here, but it is always advisable to consult legal counsel before embarking on any intercept operation. Encryption will also not be addressed.

Network forensics is the interception of computer-to-computer communication usually in real time; however, analysis is rarely performed in real time. To intercept this traffic, a tap needs to be installed. The term *tap* comes from the old telco (telephone company) days and means Test Access Port. Today the term is used more generically as a way to intercept network traffic. The actual tapping can be created several ways: hub (old school but can be effective), tap device, SPAN (this stands for switch port analyzer), or an inline device (essentially a computer with multiple network cards). These techniques require a storage device where the network traffic is collected.

The collection may be full content, meaning all of the traffic is captured, or it may just be metadata otherwise known as NetFlow[5] or IPFIX.[6] NetFlow was developed by Cisco; IPFIX is the open-source version of flow collection, and it stands for IP Flow Information Export. With enough NetFlow data, it is possible to predict behavior and effectively draw conclusions on events.

Some popular packet collection software tools used today include Wireshark, tcpdump, and Omnipeek. Tcpdump is native to just about all versions of Linux, and Wireshark is GNU public licensed software available for Linux, Mac, and Windows. Omnipeek is a commercial product.

With 50 percent of Internet traffic being encrypted,[7] full content collection may not suit your situation, so Netflow collection is the next best thing. With Flow data, although we cannot see content, we can look at the connection information and apply analytics to make a determination on what is occurring. Flow data includes source and destination IP and port addresses, date and time stamp, protocol used, the amount of data exchanged (bytes and total packets) in that particular session, and the state. Full content collection tools include Wireshark, NetworkMiner, NetWitness Investigator, Microsoft Message Analyzer, Arkime (Moloch), and many others.

CLOUD FORENSICS—WE ALL USE THE CLOUD WHETHER OR NOT WE THINK WE DO

The cloud is one of those terms that people continually banter around. Are you in the cloud? Well, the fact is that we are all in the cloud. If you are using email, shopping online, using any social media app, or storing files remotely, you are in the cloud. The cloud has become such an issue, that in March 2018, the Cloud Act was signed into law by President Trump.[8] The legal and digital forensic implications of the cloud are not trivial. Traditionally, a law enforcement officer will apply for and receive authorization from a judge through the warrant process to execute a search, at a specific location, looking for specific evidentiary items. The cloud clouds this. Where are these evidentiary items being stored? Are they even in the United States?

Some of the main players in the cloud today are Amazon Web Services (AWS), Google, Microsoft, Apple, and many others, with AWS, Microsoft, and Google collectively controlling a majority of the market, with AWS leading the Big Three.[9]

Challenges in the cloud today vary considering that every organization does things differently. Microsoft Azure cloud platform is different than AWS. Also, when we're talking cloud, are we talking about an application (e.g., Instagram); a virtual machine,[10] aka VM (instance in AWS); hardware (iron) based system; or a platform (e.g., Blackboard)? Is the cloud environment 100 percent in the cloud or is a portion of the system part of the organization (hybrid cloud)? Each of these cloud variants have their challenges. Where are these systems physically located? If they are VMs, can they be or have they been moved from one data center to another? Can you collect the data in a forensically sound manner remotely or do you need access to a data center that stores the VM? Do you require backend cloud provider logging data?

The Cloud Act allows law enforcement, with the appropriate court orders authorized by a judge of competent jurisdiction, access to data outside the United States with certain restrictions. Those restrictions include whether the subject is a U.S. person and if disclosure creates a material risk to the cloud provider for violating foreign law.[11]

Who testifies to what? The cloud provider is an innocent third party providing a technical business service. With 1.2 billion active Gmail users as of 2018,[12] the odds of a subject or a victim having an active Gmail, Outlook, or Yahoo account is high. Is cloud storage being used? Google Docs, Drop Box, iCloud, and One Drive are a few cloud storage solutions available. Apple defaults their users to iCloud, and Microsoft is making it very easy for Windows users to access One Drive. Is the storage encrypted? Who maintains the encryption keys?

Digital forensic investigators need to understand the cloud from a legal and technical perspective, which includes what data (evidence) to ask for, how to ask for it, and how to preserve it in a forensically sound manner.

MEMORY FORENSICS

The collection, analysis, and reporting on data existing in memory has become essential with modern computer systems because not all data is permanently stored or nonvolatile in computers today. Some data is only stored in memory and memory is not permanent or volatile, meaning that when the computer is turned off, that data disappears. Some malware only exists in memory; encryption keys may exist only in memory; user credentials may only exist in memory.

Encrypted file systems essentially make all data volatile. What I mean by this is if the data is encrypted, it's unobtainable. If it's unobtainable, we can call it volatile.

So what does memory forensics entail? First the examiner needs to successfully capture or image the system's memory. There are several tools available for examiners to use to capture memory. Some of these tools include Dumpit, FTK Imager, LiME (Linux), Magnet Ram Capture, Memoryze, and Volatility. All of these tools have been used effectively, but it is strongly recommended that whatever tool you use, you extensively test it on a system that is running the same operating system version and hardware as the target system if feasible. Memory imaging can fail, and if it fails, you could end up with nothing.

Once you've properly imaged memory, then you need to parse the image, identifying running applications, processes and services, and open and listing ports. As of the writing of this book, Volatility was the top forensic memory analysis tool. The FireEye (formerly Mandiant) tool Redline can be effective. Redline is not as versatile as Volatility, but its graphical interface makes it easier to use. Even extracting Ascii and Unicode[13] strings can be analytically effective.

The ability to image memory requires that the system be running, and you have the proper credentials to access the system to run your memory imaging software.

As a digital forensic examiner, memory imaging will almost always occur in the field at a target search or victim site. Memory imaging is also a onetime event. This does not mean that you can't image memory multiple times, but each image will be different since the computer is running continually, reading from and writing to memory. You will be required to identify the operating system running, deploy the proper memory imaging tool, and capture the computer's memory in the forensically sound manner properly documenting and hashing your results.

Back in the lab, you will subject the image to analysis and report your results, but it all starts with the original memory image.

MOBILE DEVICE FORENSICS

With nearly 300 million cell phones (mobile devices) in the United States[14] and 5.11 billion smartphones worldwide,[15] the odds of a mobile device being part of a criminal investigation is very high. While Apple iOS and Android devices dominate the market, there are huge number of other mobile devices used throughout the United States and the world. Apple has 1.4 billion devices worldwide,[16] and Android has 2.5 billion devices worldwide.[17] So it's safe to say that mobile devices are ubiquitous in our culture. From a forensic perspective, law enforcement must contend with a wide variety of devices in the marketplace, the operating systems they use, and the applications that run on these operating systems.

Add to the challenge that many of these phones are encrypted. Apple has been encrypting iOS by default for some time, and Android vendors have been encrypting devices by default as well.[18] All modern Android devices are encrypted by default.

You may have heard of the term "Going Dark." Going Dark was first coined by the FBI and describes the challenges that law enforcement encounters when devices are encrypted, and law enforcement has a valid legal reason for obtaining and decrypting this information but is unable to do so.

When talking about mobile devices, we should also differentiate smartphones from feature phones. Smartphones are able to download new applications; iPhones and Android phones fall into this category. Feature phones come out of the box fixed with a set of features or apps that generally can't be modified. I use the term *mobile device* to include feature phones, smartphones, and tablets.

How do you extract data from a mobile device?

- imaging
- JTAG/ISP
- chip-off

Extracting data from a mobile device is essentially an exercise in exploits, meaning the examiner uses tools that attempt to exploit the device's

MOBILedit®
by
compelson

Figure 3.2.   MOBILedit.
MOBILedit by Compelson

weaknesses to collect data. For example, in September 2019, Axi0mX published his Checkm8 exploit that can bootrom jailbreak iPhones using the A-11 or earlier chipsets.[19] The holy grail of mobile device forensics is an image, meaning a bit-for-bit exact copy of what is on the device. Getting an image may not be possible. A logical copy may be obtainable or maybe only certain data sets on the device. A logical copy gives you files but does not provide the ability to resurrect deleted data. Tools that are available to conduct this type of extraction include MSAB XRY, MOBILedit Forensic Express, Cellebrite, and Paraben.

In some situations, the mobile device needs to be physically opened (which can be a challenge itself), with data extracted via JTAG/ISP or Chip-off. JTAG/ISP means Joint Test Access Group or In-Service Programmable. With JTAG/ISP wires need to be soldered to specific pins on the mobile device board and connected to a device (e.g., RIFF Box) that can read the data stream. Chip-off involves removing a chip or chips from the board, cleaning the excess solder from the chip, and attaching the chip to a specialized chip reader. Encryption invalidates all of these collection procedures.

REVERSE ENGINEERING

Reverse engineering or RE is the most technical of all these swim lanes. RE can be broken down into three categories: static analysis, dynamic analysis, and decompilation.

Decompilation, the most technical of these disciplines, is taking an executable program (unreadable 1s and 0s) and converting it to assembler or possibly C, which is human readable but barely. Assembler is a low-level programming language that directly maps to executable operation code. C is a higher-level programming language, but both Assembler and C are difficult to read and interpret. This decompiled code is reviewed in a tool such as IDA/Hex-Rays or Ghidra. Decompilation analysis requires low-level programming skills.

Static analysis[20] is examining the code without running the code or decompiling the code. If you can't run the code or you can't decompile the code, what can you do? The first thing to do is to extract any human readable characters from the code (Ascii and/or Unicode) using a simple strings program. Either the native Linux strings command or the Microsoft Technet strings command can be used. By examining these human readable characters, it may be possible to get a feel for what the program does. We need to be careful here since it would not be difficult for a malware writer to sprinkle misleading code misdirecting analysis. Another analysis technique is to examine the code's header. In the world of Windows, it's called the PE (Portable Executable) header, in Linux its ELF (Executable and Linkable Format), and for Mac OS it's Mach-O (Mach Object File Format). So what's in these headers that is so interesting? Well, many things, but probably the most important aspect of the headers is the Import (IAT) and Export (EAT) address tables. These tables identify the other code dependencies that the program requires to run (IAT) and what resources the code provides to other programs (EAT). By looking at these dependencies, we can get a feel for what the code does.

Dynamic analysis is running the unknown code and seeing what the code does. Well, not quite. We are going to run the code, but we want to run it in an environment that is excluded from anything that could be harmed and where we can capture data (sensor placement) to analyze what the code is doing.

What Baseline Skills Do I Need to Bring?

The world of digital forensics is highly technical, but do you need a degree in computer engineering (CPE) or computer science (CS) to succeed? The short answer is no, but if you have a CS or CPE degree, that's a great start. If you've identified an agency that you wish to work for, then find out what that agency requires. It will vary. If you don't have a CPE or CS degree, where do you start in terms of baseline skills?

Table 3.1.   Digital Forensics Swim Lanes

Swim Lane	Description	Evidence Source
Digital Media	Desktops, laptops, and any media that is connected to a computer fall into this category.	Hard drives, SSD, thumb drives and other removable devices, CDs, DVDs
Network	Network traffic can be intercepted, re-sessionized, and analyzed. Sometimes only metadata is collected (aka Netflow or IPFIX).	Intercepted traffic in either pcap or Netflow/IPFIX format
Cloud	If it's stored on someone else's computer, the data needs to accessed, duplicated in a forensically sound manner, and analyzed.	VM image extracted from a cloud provider
Memory	Highly volatile and time sensitive, computer memory can be imaged and analyzed.	Memory capture
Mobile Devices	Your iPhone and Android devices need to be accessed with data forensically extracted.	Captures from Mobile devices (e.g., iPhones or Android-based devices)
Reverse Engineering	Process of examining executable code in order to determine what the code does.	Unknown or weaponized code

PROGRAMMING

Although this does vary agency by agency, there are areas of expertise where you will need to show competency. Computer programming languages are the first thing people think of when anything technical is mentioned. Programming languages take all sorts of forms. You have compiled languages like C or C++; you have interpreted or scripting-based languages like Python, PowerShell, Bash, JavaScript, and so on; you have hybrid languages like Java. Languages such as C or C++ require that the code be passed through a compiler (compiling) where an executable program is created. Scripting languages such as Python compile or interpret on the fly, meaning that the individual lines pass through an interpreter that converts the text-based code to executable code. This happens every time you want to run the program. Hybrid languages like Java have a compiled component and an interpreted component. Compiled languages execute the fastest, and interpreted languages execute the slowest. So if performance is critical, go with compiled code.

Digital forensics examiners are not programmers, but they need to understand programming and may be required to put a few lines of code together. Python is the preferred language of digital forensics, and the reason why Python is the preferred language is that there are predeveloped modules that you can call directly into your code that perform forensic activities. So if you are looking for a language to take, Python is you best option. There are also excellent development environments like PyCharm, Spyder, or Jupyter Notebook that you can use that facilitate code development.

This simple but powerful six-line program decodes an image that was coded in base64.[21] The reason why this program is so powerful yet short is because of the import base64 module, which already has all the code necessary to convert the image.

Python Example

```
import base64, os, sys

# Enter the file = "nicki.txt" you want decode from base64
print ("Base64 Decoder")
file = input("Enter the name of the file you need decoded: ")

# Enter the file name image = "nicki.jpg" of the image file
image = input("Enter the name of decoded file: ")

# Now to decode
base64.decode (open(file), open(image, "wb"))

print ("%s was created." % (image))
```

OPERATING SYSTEMS

Computers need an operating system to run. The three most popular operating systems today are Windows, Mac OS, and Linux. You will invariably come in contact with computers that run one or more of these systems. Knowledge of the command line for operating systems is essential, as well as knowledge of the file systems these operating systems use.

In the world of desktops and laptops, Windows is king with 78 percent of the market. MacOS has about 14 percent of the market and Linux about 2 percent,[22] but these numbers are deceiving. When you look at the enterprise server and cloud deployment markets, Linux is significant with 70 percent of the market, and this percentage is probably low. The reason why Linux is so popular at the enterprise and cloud level is because it's essentially free, powerful, and fast.

Windows

For Windows, as an examiner, operating at the command line is crucial for several reason. First, hackers use the command line. Second, command line programs generally take up fewer computer resources. Third, not all functionality can be had from the graphical user interface.

There are two command line environments in Windows: cmd.exe and PowerShell. Cmd.exe (formerly command.com) has been around since the DOS[23] days.

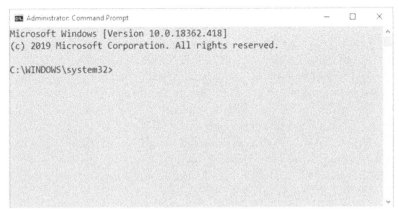

Figure 3.3.   Windows command line. *Figure created by authors*

PowerShell has been around since about 2006 and is relatively unknown by lay users, but hackers know PowerShell very well. At first glance, there doesn't seem to be much difference between the standard Windows command line and PowerShell, but appearances can be deceptive. SecurityWeek reported that PowerShell is the most prevalent hacker technique being detected and is being detected twice as often as the next most common technique.[24] PowerShell allows you to get completely under the hood of Windows because it has all the functionality that is provided by .NET.[25]

Figure 3.4.   PowerShell command line. *Figure created by authors*

Legacy cmd.exe commands usually run on PowerShell, but the PowerShell version of these commands offers much more versatility.

Windows needs a file system, a way to organize the stuff that the operating system has access to. There are several file systems that Windows can access, but its main file system is NTFS. NTFS stands for New Technology File System and was first released in 1993,[26] and NTFS in its current release hasn't changed much in the last eighteen years. Other file systems that Windows can use fall under the category of FAT systems. FAT stands for File Allocation Table. The most recent version of FAT is call ex-FAT and was developed by Microsoft as a lightweight file system that cameras and other similar devices could use. ex-FAT's predecessors are FAT-32, FAT-16, and FAT-12. FAT-32 is still used, but you don't see much FAT-16 and FAT-12 anymore.

Linux

Linux was released by Linus Torvalds in 1994 as he wished to use a Unix type operating system.[27] Linux is open source, meaning that you can use it for free. Linux has morphed over the years with countless different distributions (distros) of Linux. A few of the more well-known distros of Linux are: Red Hat Enterprise Linux (RHEL), Fedora, Ubuntu, Debian, and countless others. There are also forensics versions of Linux to include Kali[28] and Paladin.[29] Both are excellent platforms with Kali better suited for the pen test environment and Paladin digital forensics.

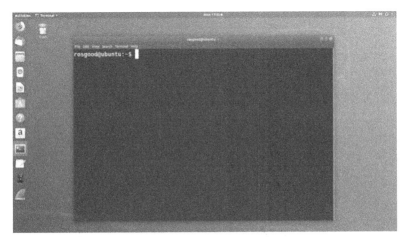

Figure 3.5.   Ubuntu Linux. *Figure created by authors*

Linux has several command shells that are available, but the most popular command shell is bash or the Bourne Again Shell, which is the successor to the Bourne Shell. There is also the Korn Shell (ksh) and the C Shell (csh). Bash is native to just about all distros of Linux.

Linux has the ability to integrate many files systems into its native ext virtual file system. The latest version of ext is ext-4. Linux can also integrate NTFS and FAT-32. Windows can't access an ext file system.

Mac OS

Mac OS is Apple's operating system for its desktop and laptop computers. Mac OS has been around since 1984 but has undergone significant changes.

The latest version of Mac OS is Catalina. Mac OS has a Linux-like command shell. In fact, if you know bash, you can operate the Mac OS command shell. Modern versions of Mac OS use Apple File System (APFS). Mac OS will also interact with FAT and ex-FAT.[30] Mac OS is based on Carnegie Mellon's Mach OS kernel; Mach OS is based on the BSD (version of Unix) kernel, but it is not Unix/Linux. Mac OS has many features unique to Apple so don't make the mistake that it's just another flavor of Linux.

NETWORKING

Everything today is connected to the internet. Well not everything, but those things not connected to the internet are few and far between. Your laptop is connected, your phone is connected, your tablet is connected, your TV is connected, and probably your toaster and refrigerator as well.

The eight-hundred-pound gorilla in the world of networking is TCP/IP (Transmission Control Protocol/Internet Protocol). Your local home network uses TCP/IP, your school and business use TCP/IP, and the government uses TCP/IP. TCP/IP is not a protocol but a series or suite of protocols that interact together to allow for seamless network connectivity.

Table 3.2. TCP/IP Stack

Layer	Name	Description	Identifier
5	Application	Programs that utilize network resources (e.g., browser, email client, ping, etc.)	Application Name
4	Transport	Application-to-application/ session-to-session communications	Port Number
3	Network	Intersegment or network-to-network communications	IP Address
2	Datalink	Intrasegment (LAN) communication	MAC Address
1	Physical	fiber, cable, or radio-based connection (cellular or 802.11 Wi-Fi)	Cable Pair/Binding Post or Frequency

Knowledge of the intricacies of the TCP/IP stack is an essential prerequisite for examiners, especially how the stack layers interact with one another. A complete description of TCP/IP is a book[31] in itself. There are a few items that need to identified:

- Layer 1

  ○ Cabling may require the deployment of fiber optic cable or CAT 5/6 cable and the termination of that medium.
  ○ Wi-Fi or cell tower signals may need to be blocked from a device to preserve evidence.

- Layer 2

  ○ In a cabled network, you will most often be dealing with ethernet.
  ○ In a wireless environment, you will generally be dealing with 802.11 or Wi-Fi.

- Layer 3

  ○ This is home of the IP address and there are two versions used today: IPv4 and IPv6.

    ■ IPv4 is a 32-bit address often expressed as a decimal number.

      • IPv4—15.21.33.123

    ■ IPv6 is a 128-bit address that often expresses as a hexadecimal (base 16) number.

      • IPv6—2002:0011:4c65:d124:6c24:10ed:4c89:3b58

    ■ Some IPv4 and IPv6 addresses are for internal use only and not routable on the internet.

      • 192.168.0.0—192.168.255.255
      • 172.16.0.0—172.31.255.255
      • 10.0.0.0.- 10.255.255.255
      • FE80::

    ■ Identifies the computer on both sides of the communications channel (e.g., server and client)

- Layer 4

  - Sessions are identified by port number, and these port numbers point to the application providing the service (e.g., port 443 denotes a web server).
  - Some port numbers are reserved for particular applications and some are not reserved.
  - The source IP address: source port number: destination IP address: destination port number creates a globally unique address called a socket.
  - Port can be active (data is being transmitted), passive (waiting to communicate with someone), or closed.

- Layer 5

  - There are many applications that use network resources.
  - Malware, or weaponized code that uses network resources, is a type of network application, meaning they need port numbers and IP addresses to communicate.

Data gets from one computer to another through a series of switching and routing maneuvers by switches and routers. Generally speaking, switches[32] permit data to move within a network and routers permit data to move between networks, but there are exceptions to this (e.g., MPLS).[33]

ADVANCED SKILLS

For those that are interested in malware analysis, especially code decompilation analysis, you will need a thorough understanding of Assembler and C programming languages as well as a knowledge of compilers. Decompiling executable code (program) will provide you with Assembler and/or C code. This is a specialty within digital forensics possessed by a limited group.

SOFT SKILLS

So the hard skills required by digital forensics examiners are extensive but obtainable. But there are soft skills that may be required that are often overlooked. These soft skills include interview or human interaction skills, report writing or written communications skills, and analytical skills.

Written Communications Skills

All examiners are required to write reports. These reports may end up in a court of law (criminal or civil) or be provided to executive management. Having your written work reviewed in a court of law is the ultimate adversarial peer review. Every comma and bolded semicolon will be dissected, interpreted, misinterpreted, and subject to both direct examination and cross-examination.

Because of the volume of digital evidence in a case, digital forensics tools often incorporate report-generating tools to facilitate the reporting requirement. You are responsible for every word in that report. The excuse "I don't know; that's what Axiom, Blacklight, or Truxton produced" could very well lose the case for the prosecutor, especially if the critical evidence is digital.

Agencies should have standard report templates that are required to facilitate report preparation as well as a peer review process before the report is finalized to catch glaring errors. The use of standard grammatically correct English is a must with the avoidance of as much techno slang as possible. When techno slang is used, it needs to be explained or be explainable in such a way that the average intelligent juror, attorney, and judge can understand. The last thing you want in a courtroom is a judge chastising you for not being understandable.

These reports must be demonstrably correct. Even an honest mistake can derail a prosecution. At best, the crossing attorney will force you to acknowledge the errors, which may have a significant impact on the jury. The writing style needs to be structured, to the point, without embellishment. Remember, we are writing an informative, factually correct, focused document, not the great American novel. Having said that, digital forensic reports are often voluminous.

Interviewing Skills

Some agencies do not allow their examiners to interact with the public; some do. For those that do, your ability to effectively engage a victim, subject, or involved third party can make or break your case.

Stop and Shop Case

In the late 1990s, there was a federal investigation targeting Stop and Shop, a large grocery store chain, for defrauding vendors. The details of the fraud revolved around how grocery stores get paid or reimbursed when vendors offer sales or discounts. When a manufacturer, say BobCo, wants to put canned peas on sale, they will notify the grocery store and the grocery store will place that item on sale. That doesn't sound fair because the grocery store will have to absorb the sales price reduction.

In reality, the grocery store reports back to the manufacturer how many cans of peas were sold, and the manufacturer reimburses the grocery store for the sales price reduction for each can of peas sold.

What Stop and Shop was doing, and they were doing it with computer programming, was increasing the number of cans of peas sold and reported to the vendors by 20 percent or 1.2. This doesn't sound like much, but multiply this by every sale by every manufacturer that Stop and Shop deals with, and you now have some real money. A can of peas here, a sack of potatoes there, and pretty soon it adds up to real money.

My job, on the morning that a search warrant was to be executed on Stop and Shop headquarters in Massachusetts, was to visit the programmer that coded the fraudulent entries and get that programmer to cooperate. Why was an examiner assigned to this interview? Because the examiner understood programming.

After overcoming the programmer's fear— of two people with guns coming to the house at 6:00 a.m.—we were able to establish a rapport. That rapport led us to identifying the program that was facilitating the fraud and exactly where in the Stop and Shop headquarters the program was stored.

We joined the search team at the Stop and Shop headquarters, where we met the Case Agent and the Stop and Shop attorneys. We walked directly to the source code library, identified the source code, went to actual lines in the code, and identified where Stop and Shop was fraudulently inflating it sales.

This evidence was critical to the speedy resolution of this case. What this took was a little programming skill—COBOL to be precise—and some good people skills. Without the help of the programmer, the odds of identifying this critical piece of evidence would have been slim to none.

As examiner, you could interview a systems administrator at a data center, a victim of an online extortion, the subject in a child exploitation case, or anyone else for that matter. Generally, the focus of your interview will be to facilitate the collection of digital evidence, but you never know where the interview will go.

People that are technically driven are often not the best people persons. But with a little practice, rapport building and solid interview skills can be a tool on your toolbelt.

There's an old NSA joke: how can you tell an introvert from an extrovert at NSA? Answer: the extrovert is looking at other people's shoes. The ability to make eye contact, shake hands, establish a rapport, and get people to trust you is worth more to the forensics evidence collection process than can be described here.[34]

LEGAL SKILLS

Most if not all law enforcement agencies provide some legal training to their examiners, but it doesn't hurt to be a little prepared since what you do as an examiner will have legal consequences. The goal here is not to turn you into an attorney but to make you legal aware. Many potential legal problem can be avoided with a little prior planning—Prior Planning Prevents Piss Poor Performance (the six Ps).

Chain of Custody

In the world of law enforcement, the ability to prove that a piece of evidence has not changed from the time it was originally acquired until it is ultimately no longer required (disposed of) is critical to the prosecutorial process. This is known as chain of custody (CoC), and CoC has its challenges and requirements in the world of digital evidence, especially since in the world of ones and zeros, the ability to alter or contaminate evidence is of definite concern.

Prior to the proliferation of mobile devices and the use of encryption, most digital forensic acquisition was accomplished by starting with an unpowered device, attaching a write-blocker[35] to it, powering the device, then extracting a bit-for-bit image of the device that can be verified through cryptographic hash. Today, although this technique is still in use, it often becomes necessary to extract evidence from a live device to overcome the encryption challenge or, as with mobile devices, the device must be on to extract data.

Maintaining detailed notes or records of what you do is critical to the process. If you conduct an exam and then three years later are required to testify to your exam results, your notes are the only way you are going to remember what you did. Your notes need to be detailed enough that another examiner can reproduce your work from them.

The cryptographic hash (MD-5, SHA-X[36]) are the digital fingerprints of electronic evidence verification. Just about all digital forensic tools provide hashing. They have to be digital forensic tools.

Manufacturing of Digital Evidence in Turkish Military Coup Case

In 2010, Turkish authorities seized a hard drive from the Turkish Naval Command, and the evidence obtained from this hard drive was key evidence in the arrest and conviction of Turkish military officers. In 2013, at the request of the families of the imprisoned military officers, Mark Spencer and his team at Arsenal Consulting received a copy of this drive. Using a technique developed by Arsenal called Anchor Relative Time, it was determined that the evidence obtained from this hard drive had been planted. As a result of digital forensic work conducted by Arsenal, innocent people were cleared of the crimes accused.

By comparing log entries from $LogFile, log sequence numbers, from the hard drives file system (NTFS), Arsenal was able to determine that the evidentiary digital files could not have been on the drive prior to its seizure, meaning the evidence was planted after the drive was seized by the Turkish government.

The takeaway from this sobering case is that digital evidence can be fabricated, falsified, contaminated, and/or destroyed with relative ease, so following chain of custody collection and analysis procedures is critical to the digital forensics process. As digital forensics examiners, it's absolutely essential that we be objective. Although we are law enforcement, we treat every case from a neutral perspective.

Source: Mark Spencer, "Beyond Timelines—Anchors in Relative Time," *Digital Forensics Magazine*, no. 18 (February 2014): 15–19.

Other Legal Stuff

Without having to graduate from law school, there are a few more concepts with which we should be familiar. Court orders, warrants, subpoenas, and consent are all legal concepts that the digital forensics examiner needs to be familiar with since they all deal with evidence collection.

Subpoenas

Subpoenas are legal compulsion requirements issued through a grand jury by a prosecutor. Evidence obtained through subpoena is the property of the grand jury and is subject to the rules and secrecy of the grand jury. Improperly divulging grand jury material can be a violation of law (e.g., Federal Rule 6e). At the federal level, some agencies such as the FBI and DEA have some limited administrative subpoena (or "subpena" as some government agencies spell it) power used primarily to collect information such as phone billing and call records.

Court Order

A court order is a prosecutorial demand authorized by a judge to require some entity to produce something of potentially evidentiary value. The judge usually requires some fact-based explanation in the form of an affidavit as to why the government wants this information.

FBI Gets a Court Order Compelling Apple to Cooperate in San Bernardino Terrorism Massacre Case

One of the most famous digital forensics court orders was the February 2016 court order obtained by the FBI from Judge Sheri Pym, Central District of California, ordering Apple Computer to assist the FBI access the iPhone 5C by circumventing the ten try wipe feature of the iPhone's operating system. The iPhone in question was used by Syed Farook, who murdered multiple coworkers.

This court order was challenged by Apple Computer and was essentially withdrawn by the federal government when the FBI no longer required Apple's assistance.

Source: U.S. Department of Justice, "Order Compelling Apple, Inc. to Assist Agents in Search," February 16, 2016, https://www.justice.gov/usao-cdca/file/825001/download.

Search Warrant

A search warrant is a type of court order that authorizes a law enforcement officer to search a specific location and seize specific items or information. Search warrants must be accompanied by an affidavit sworn to by a law enforcement officer where that officer has articulated to the judge that probable cause exists that a crime(s) has been committed and that evidence of that crime can be found at the location identified.

Sounds straightforward, but in the world of digital forensics, this is not always the case. Consider the situation that you have a warrant to search for evidence of mortgage fraud at Mortgages R Us, 9876 Main Street, Mortgage Land, Maryland. You execute the warrant, and you observe evidence on a computer's screen, but what you are seeing is not stored locally but data stored at AWS. Extracting that data is most likely outside the scope of your warrant since your warrant only covers the Main Street address. You will probably need to get a second warrant as soon as possible. Also, the judge may only allow you to image or copy data that is at the search location instead of seizing hardware. One reason that the judge may not allow for hardware seizure is where said seizure may create a going concern issue.[37]

Consent

Law enforcement officers have a right to ask someone for permission to take/obtain evidence: "Excuse me, sir, is it OK to search your computer?" That person can consent or say yes. Should that officer find anything evidentiary, then the officer has the right to seize that evidence. Of course, the person also has the right not to consent to the search or to rescind the authorization to search at any time. This consent may be verbal, or the agency may have a consent form that the officer will have that person sign. That person may also refuse to sign the consent form but consent to the search. Such consent is still valid at the federal level; however, at the state and local level, please consult the regulations for that municipality. A third party or victim can consent, but it's also possible that a subject may consent.

NON-EXAMINER-BASED ANALYTICAL SKILLS

Digital forensics examiners may also, and often do, perform analytical work, especially given the examiner's technical expertise. Knowledge of basic analytical techniques will be critical. String searching, linkage analysis, anomaly detection, visualization, timeline analysis, and frequency analysis are just a few techniques that can be used in digital forensics investigations.

String Searching

Just as you would with Google, a string search consists of entering one or more key words into a forensic tool's search engine looking for positive hits. The longer or more specific the search term or key word is, the better the search generally. "The" is a bad word to search. "Supercalifragilisticexpialidocious" would be a great word to search. The point here is to reduce the number of false positive hits by improving search efficiency. Just about all digital forensic tools have string search capability. A word of caution though: strings come in different formats. In Windows, it's Unicode. In Linux, it's Ascii. Unicode is 16 bit and Ascii is 8 bit. This bit thing is the number of 1s and 0s assigned to one character.

Figure 3.6.   Truxton string search for "bank" revealing a spreadsheet with banking information. *Truxton Forensics (https://truxtonforensics.com/)*

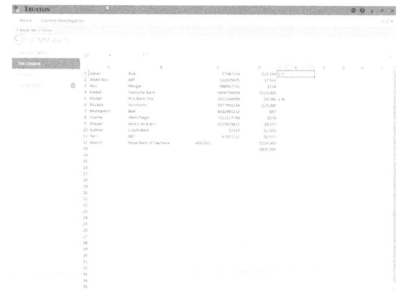

Figure 3.7.   MM.xlsx bank account information. *Figure created by authors*

Linkage Analysis

Often used in larger conspiracy cases, this technique attempts to connect one subject to another through some specific artifact or artifacts. That artifact may be a phone number, an email address, a Twitter handle, a location, a cryptographic hash, or any combination thereof. The purpose of linkage analysis is to establish the existence of relationships. Linkage can be direct— Bob called Lucy via 555-1212—or linkage can be indirect—Bob called Lucy via 555-1212 and Lucy contacted Esmerelda via emelda44@mail.com. By establishing these relationships, we can focus our investigation on the right people. Digital forensics provides a plethora of artifacts that can link people.

Anomaly Detection

We are looking at what is out of the ordinary. Why did Erasmus log into his office computer at 2:00 a.m. on a Saturday when he normally leaves work by 5:00 p.m. and doesn't work weekends? Why was there a transfer of 8 gigabytes of data from our file server to 27.50.41.125 on October 21, 2019, at 11:34 p.m.?

For anomaly detection to be effective, you need to know what is normal for that system. Baselining a system or network can be very valuable since you now have an idea of what's normal, making it easier to figure out what is "abby normal" (in the parlance of the movie *Young Frankenstein*).

Visualization

One picture can tell a thousand words. The ability to visually represent data in a chart, graph, diagram, and so on can be of great value to an investigation and its subsequent successful prosecution.

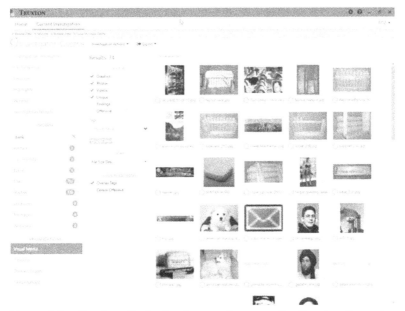

Figure 3.8.   Truxton displaying images from media. *Truxton Forensics (https://truxtonforensics.com/)*

Timeline Analysis

Placing events on a timeline can reveal relationships to people, places, and events that may not be readily obvious. Often, evidence is represented in a timeline in a court of law to bring perspective to the evidence. When a subject logged in to a computer, when a file was downloaded, and when a program was run can link an individual to an act.

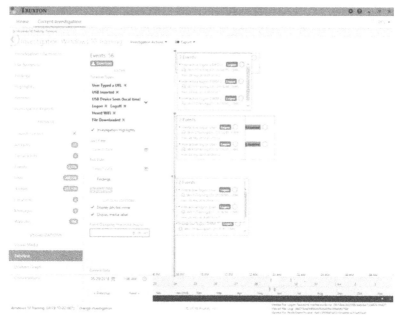

Figure 3.9. Truxton timeline analysis. *Truxton Forensics (https://truxtonforensics.com/)*

Frequency Analysis

Frequency analysis is in some way the opposite of timeline analysis. In frequency analysis, you are counting or recording how many times something occurs or its frequency. How many times did Imelda call 212-356-0243? How many times did Vladimir's computer connect to 27.50.42.101?

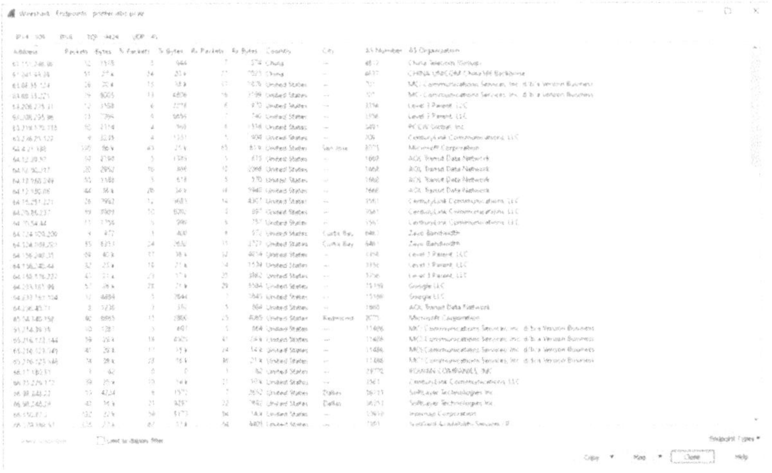

Figure 3.10.   Wireshark IP address analysis. *Figure created by authors*

Summary

Digital forensics examiners need a wide variety of technical, analytical, legal, and human relationship skills, making a career in digital forensics truly multidisciplinary. The better your skill set, the more employable you are. Determining what specific skills an agency requires will allow you to focus your education, training, and experience.

CHAPTER 4

Education and Certifications

Cybersecurity and Digital Forensics Programs

Not all cybersecurity/digital forensics programs are alike. The challenge for those of you looking for cybersecurity/digital forensics education is how your education best fits your own personal situation. Does an online program work best or are you looking for a brick-and-mortar (in-class) program? Should you look at certification programs? Do you already have an undergraduate degree? If so, does a master's program work for you? Do you already work full-time so a part-time program would be best for you?

If you are just starting college and are technically inclined, then computer engineering or computer science programs are generally a solid place to start. That sounds rather mundane, but this background sets you up perfectly for the career in digital forensics and cyber analysis.

So what's the difference between computer engineering and computer science? Good question. When high school seniors ask me this question, I usually respond with "Do you want to be a coder or a burner?" I usually get "What does that mean?" Although the difference between computer engineering and computer science degrees often blurs today, the fundamental difference is that computer engineering is primarily focused on hardware design: chips, motherboards, sensors, systems, and so on. Some computer engineer designed the iPhone, Android, your laptop, and everything else with a computer chip in it. Computer science focuses on programming. A good computer science program should graduate students that have an in-depth knowledge of operating systems, compiler design, and device driver programming.

What about database design, website development, and general application programming? Computer engineers and scientists can do that as well, but many universities have information technology programs not quite as rigorous as computer engineering or computer science that fulfill these skills. Most computer engineering/science programs require four semesters or more of calculus and four semesters of physics, whereas information technology programs are not as math and physics heavy.

A successful digital forensics/cyber analysis career can be achieved through a formal education, informal training through certifications, or a combination of both. We also made the case for the important connection between digital forensics/cybersecurity and criminal justice. This chapter will expand on how a student can approach their digital forensics career by using the formal or training track, depending on what their career goals are.

Deciding whether to get an associate's (two-year), undergraduate (four-year), or master's degree can prove to be challenging. However, once this hurdle has been cleared, it gets easier. The first step a student needs to decide is what type of digital forensics/cybersecurity career they want. It goes without saying that if a student wants a long-term career in digital forensics, they will need to invest in a formal academic education, training, or a combination of both.

What's the difference between cybersecurity and digital forensics? If you look at these two highly related disciplines from a Venn diagram[1] perspective, both overlap but include features that the other does not.

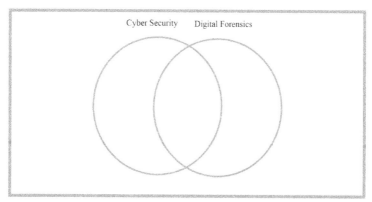

Figure 4.1. The intersection of cybersecurity and digital forensics. *Figure created by authors*

The intersection of cybersecurity and digital forensics focuses primarily, but not exclusively, on incident response. Digital forensics is primarily offensive in nature: seizing laptops, phones, and media; intercepting network traffic; and so forth. Cybersecurity is primarily defensive in nature, although there are components that are offensive. These offensive components exist at the federal government level since it is a violation of the Computer Fraud and Abuse Act to hack or even hack back. Penetration testing, the use of tools, techniques, and procedures with consent to test cyber defenses, is offensive in nature, but it's legal because the attacker, the pen tester, enters into an agreement with the victim, aka customer.

A significant portion of cybersecurity is process-based. These processes include but are not limited to authentication, authorization, and validation, or to put it another way: is the data in question only accessible by the right users at the right time and is that data corrected or only changed by the right users? These processes include access control, encryption, two-factor authentication, virtual private networks, and all the policies and procedures required to put those processes into effect. When a breach of cybersecurity occurs, the tools, techniques, and procedures of digital forensics take over. In the world of cybersecurity, a breach will invoke a response to some degree. This response is incident response or IR.

Digital forensics is more than just IR and includes criminal investigations, civil matters, intelligence collection and analysis, and internal inquiries. Digital forensics is generally not concerned with security per se[2] and usually requires that the examiner circumvent security measures to be effective. Digital forensics examiners often need to circumvent access control, encryption, user credentials, and so on to collect data that may be of evidentiary value.

Certificate Programs (Certs)

Certificate programs are offered by commercial entities as well as universities. Certificate programs are concentrated in nature. Universities may offer four or five courses as a certificate. These certificates may be at the undergraduate level or the graduate level. If at the graduate level, then an undergraduate degree is required.

Commercial entities will offer a course or group of courses. These courses may align with certain software packages or may be broader in scope. To obtain a specific Cert, you may be required to own a valid license to the software or hardware.

There are several certificate programs that, generally for a fee, require an exam. Some of these digital forensics certificate programs include:

- Global Information Assurance Certifications (GIAC) through SANS: https://sans.org
  - Certified Forensic Analyst (GCFA)
  - Advanced Smartphone Forensics (GASF)
  - Certified Forensic Examiner (GCFE)
  - Network Forensic Analyst (GFNA)
  - Reverse Engineering Malware (GREM)
- EnCase Certified Examiner (EnCE) through Opentext: https://www .opentext.com/products-and-solutions/services/training-and-learning -services/encase-training/examiner-certification
- AccessData Certified Examiner (ACE): https://accessdata.com/training/ computer-forensics-certification
- Certified Blacklight Examiner (CBE): https://www.blackbagtech.com/ training/certifications/certified-blacklight-examiner-cbe/

As you can see, many of these certification programs are attached to specific software: EnCE—EnCase, ACE—FTK, CBE—Blacklight. So if you are going to use a specific digital forensic software platform, it may not be a bad idea to certify in that tool. That being said, a Cert program may not be enough to show across-the-board competency as a digital forensics examiner. A good examiner can adapt to just about any tool. A good examiner is tool independent.

Prospective employers may be willing to pay for certain digital forensic Certs but may require general technology Certs before considering you. These Certs may include:

- A+
- Net+
- CCNA or some other Cisco Cert
- Security+
- Linux+

Having these Certs, or ones similar to these, does not mean that you are God's gift to technology, but it shows a level of competence far above that of the average user.

A+, Net+, Security+, and Linux+ are all offered by CompTIA (https:// www.comptia.org/certifications). The exams cost between $200 and $450 each, but that is only the cost for the exam. Training and course preparation materials vary in price. Study guides can be found online for $40 to $50 for those that do not require a formal course.

Cisco's Certified Network Associate—Routing and Switching (CCNA R&S) and other Cisco exams are offered by, of course, Cisco (https:// www.cisco.com/c/en/us/training-events/training-certifications/exams/ current-list.html). There are an almost endless number of Cisco certifications. The CCNA R&S is one of Cisco's base Certs and covers routing and switching topics that are germane to base skills required by examiners. There are a number of simulators on the market today that can help you train for base-level exams. A few of these simulators include:

- Cisco Packet Tracer: https://www.netacad.com/courses/packet-tracer ?dtid=osscdc000283
- Boson NetSim: https://www.boson.com/netsim-cisco-network-simulator
- GNS3: https://www.gns3.com/
- VIRL: http://virl.cisco.com/
- EVE-NG: https://www.eve-ng.net/

The above simulator/emulators vary in price from free to about $350 for the NetSim CCNP version, and GNS3 and EVE-NG require that you have access to router/switch operating system binaries. Simulators are a fairly easy way to get some hands-on skill, but nothing beats jacking into a console port of a router or switch. If you search online, you can find end-of-life Cisco devices that are very inexpensive.[3]

One last note on Certs: if where you want to work requires them, then get whatever Certs are required.

Formal (Academic) Education

Formal education as we define it here entails going to a two- or four-year college or university and/or getting a master's degree in digital forensics/ cybersecurity. Depending on where students see themselves, there are various approaches. First the student has to decide what specific career track they see themselves in. The most visible career option for most criminal justice students is law enforcement, for which they will need to invest in college courses like cybersecurity, digital forensics, networking, operating systems, and basic programming/scripting, just to mention a few. Depending on the college, they may need to take these classes from other departments like computer engineering, computer science, or information technology as electives or by declaring a minor in such majors.

If a student is just starting their college search, they will need to select where to get their associate's or bachelor's degree. Again, this depends on what kind of career they want. There are various types of educational institutions offering digital forensics/cybersecurity degrees. Many such degrees are different from each other, and there is a need for a student to identify what kind of institution best fits their career goals.

A student who has already graduated with a bachelor's degree can get a master's degree or take the training route, using certifications to gain the skills needed to enter the field. If a student is looking at a long-term career in management, they would benefit from a master's degree, especially if they want a digital forensics/cybersecurity consulting career. If a student wants a career in law enforcement, however, they may also need certifications in addition to formal education as required by the law enforcement agency. Certifications are a way to learn the valuable, up-to-date techniques currently being used in the field and will be discussed later in the chapter.

UNDERGRADUATE PROGRAMS

With undergraduate programs, you are essentially looking at either two- or four-year programs. With two-year programs, it's a plus if the program can roll up into a four-year program. For example, in Virginia, the community college system has formal agreements with the four-year universities that facilitate a student transitioning from a two-year to four-year program. You may also save a significant amount of money, if done correctly, by first attending a two-year school since tuition for many of these schools is significantly lower than four-year schools.

In looking at an undergraduate program, some things to consider are:

- Does the program sponsor internships?
- Is there a work-study option?

- What is the focus of the program (technical vs. policy/management)?
- Faculty industry experience.

Internships and work-study programs are a great way of actually getting some experience and can often lead to a job offer. The focus of the program is critical since, as a future examiner, you will require hard technical skills. A program that is management or policy focused may work well for certain analyst positions, but it's the hard skills that will get you the examiner job.

GRADUATE PROGRAMS

Graduate programs come in many shapes and sizes. Most offer part-time enrollment and can be completed in three years or less. If you are looking for hard digital forensics skills, then look for programs offering hard skill courses. For those with nontechnical undergraduate degrees (e.g., criminal justice), look for programs that will admit you provisionally and provide a set of additional courses that will hone your base technical skills.

Components of an Effective Digital Forensics Program

So you are looking for the right digital forensics program. What components should the program have that will make it effective?

The following list is what needs to be considered at the core of any digital forensics program:[4]

- digital media
- mobile devices
- network intercept and analysis
- reverse engineering (RE)
- incident response (IR)
- penetration testing (PT)
- memory analysis
- law and ethics
- moot court

A program that doesn't contain these core components may not contain the academic rigor required to get you to where you need to go. Also, one or more of these digital forensics areas needs to incorporate scripting to include PowerShell, BASH, and Python.

Digital media, mobile device, network intercept, RE, IR, PT, and memory analysis are the tools and techniques side of digital forensics. Law, ethics, and moot court provide the practical framework from which digital forensics works.

Moot court is where a student gets the opportunity to practice testifying. Usually, as part of a course or courses, a fact pattern is developed where a student can collect, analyze, and report on evidence. The student's digital forensics report is submitted as evidence in a mock trial. The student will take the witness stand and proceed to be questioned by the prosecution (direct testimony) and then by the defense (cross-examination). This is an adversarial situation where you as unbiased reporter of fact are to answer questions based on the facts you've identified. Moot court can be an eye-opening experience. The other participants in the proceeding can be trained faculty, attorneys, and judges that usually donate (pro bono) their time to this worthwhile learning experience. It's much better to learn from your mistakes in moot court than to repeat those same mistakes in the real world.

Law and ethics give us the purpose and boundaries that examiners live by. The United States is not just one set of laws but an amalgamation of laws from the federal, state, and local levels. Also, with the European Union (EU) General Data Protection Regulation law (GDPR) and California adopting its own version of GDPR called the California Consumer Privacy Act (CCPA), examiners need to be cognizant of the ever-shifting legal environment of the world around us. Ethics provide the moral boundaries within which we live. Just because something is legal doesn't make it the ethical thing to do. An example of this is a friend who lets you borrow his laptop for a class assignment, so legally you have been given consent to use this computer. Consent to use may legally allow you to access data on the laptop, but you do not have an ethical right to snoop around your friend's computer.

What should courses contain from a hands-on perspective?

- *Digital Media:* Drive imaging should be part of the curriculum. Ideally, students should have to extract a drive from a computer, image that drive, then reinstall the drive, returning the computer to working order. The drive could be magnetic, solid state, or solid state chipsets.

- *Mobile Devices:* Forensic extraction of data from a cellphone.
- *Network Intercept and Analysis:* Physical interception of ethernet network traffic and its analysis.
- *Reverse Engineering:* Taking unknown executable code and performing static, dynamic, and decompilation analysis.
- *Incident Response:* Taking a breach from initial alert through remediation and recovery.
- *Pen Testing:* Performing a pen test on a mock organization.

There are many other course topics that can be covered in a digital forensics program. These other topics include intrusion detection, dark web, crypto currency, Mac forensics, Linux forensics, cloud forensics, digital profiling, and fraud analysis, just to name a few.

The best way to learn digital forensics is from the people that do it for a living. What is the experience base of faculty teaching at the institution? Digital forensics is an applied endeavor, so people with hands-on experience will generally be able to impart their knowledge more effectively.

Online Programs

CyberDegrees.org (https://www.cyberdegrees.org/listings/best-online-computer-forensics-programs/) lists twenty-two online programs at the time of writing. These programs include the following:

1. Boston University Metropolitan College (graduate)
2. Pace University New York (undergraduate)
3. Utica College (undergraduate and graduate)
4. University of Rhode Island (graduate certificate)
5. Champlain College (undergraduate and graduate)
6. Middle Georgia State University (undergraduate and graduate)
7. University of South Florida Main Campus (graduate)
8. Penn State University Main Campus (graduate)
9. University of Central Florida (graduate)
10. University of Maryland Global Campus[5] (graduate)
11. University of the Cumberlands (graduate)
12. Robert Morris University (undergraduate)
13. Sam Houston State University[6] (graduate)
14. Norwich University (graduate)

15. Old Dominion University (undergraduate)
16. Georgia Southern (graduate)
17. Fairleigh Dickinson University (graduate)
18. Stevenson University (graduate)
19. DeSales University (graduate)
20. University of Colorado–Denver (graduate)
21. University of New Haven (graduate)
22. Oklahoma State University Institute of Technology–Okmulgee

When looking for an online program, it may be a good idea to see if that institution also has a brick-and-mortar program. This shows a level of commitment to digital forensics over and above online offerings. Sam Houston State, University of Central Florida, and Champlain College offer in-class or online programs in digital forensics. Is the program synchronous or asynchronous? Synchronous is where the class meets at a specific date and time online, and asynchronous is where the class has no formal meeting time. There are also programs that are a mix of synchronous and asynchronous, and some may also have an in-person on campus requirement.

Cost

Everything costs money. If it's free, well, you get what you pay for. Digital forensics education and training costs vary. Two-year programs can run from $20,000 to $30,000. Four-year programs can run from $75,000 to over $200,000. Graduate programs can run from $21,000 to $50,000. Certification exams and the courses that support those exams vary wildly from free to thousands of dollars.

For the cost-conscious undergrad, combining two-year and four-year programs where available is generally the most cost effective if you can roll your two-year program directly into its four-year program counterpart.

How to Pick an Institution

So you are ready to commit to a digital forensics program. To whom do you apply and what are your chances of acceptance and successful completion?[7]

Table 4.1. Cyber Security and Digital Forensics Intersection

	Yes	No
Cert Programs		
Are Certs required?		
What Certs are required?		
Are they Certs that will get you the job?		
What is the cost of Cert?		
Will you need to pay for the software/hardware covered by the Cert?		
College Programs		
In-class?		
Online?		
Two-year program?		
Four-year program?		
Certificate?		
Graduate program?		
Will you need foundation or prerequisite courses?		
Does the college offer the right mix of courses?		
Does the college have the right faculty teaching?		
Is the program primarily hands-on?		
Will you get access to equipment in labs?		
Financing		
Employer?		
Veterans?		
Loans?		
Pay as you go?		
Both College and Certs		
Which will provide the greatest impact sooner?		

Note: Table created by the authors. Do not confuse college certificates with commercial certificate programs. Many colleges offer abbreviated programs called certificates. The number of credits will vary but is usually between twelve and eighteen credits. These are not college degrees, but you will usually be taking the same courses that are in the larger degree programs.

THE CENTERS FOR ACADEMIC EXCELLENCE (CAE) DESIGNATED INSTITUTIONS

Another decision that is important here is whether to get a degree from a Center for Academic Excellence (CAE) or non-CAE institution. While the CAE designation is an important accreditation for an academic institution, It Is not entirely necessary to attend one to have a rewarding digital forensics or cyber analysis career.

The CAE designation is awarded through a joint program by the Department of Homeland Security (DHS) and the National Security Agency (NSA) to qualifying educational institutions willing to teach and do research in cybersecurity. The NSA started the CAE program in 1988 and was joined by the DHS in 2004 with the recognition that cybersecurity lapses were an emerging threat to critical infrastructure. The designations are awarded yearly to institutions who apply and are given after a rigorous evaluation.

The CAE program has three designations: Centers of Academic Excellence in Cyber Defense Two-Year Education (CAE/2Y), given to two-year colleges such as community colleges; Centers of Academic Excellence in Cyber Defense (CAE/CDE), given to four-year institutions with a focus on information assurance (IA); and Centers of Academic Excellence in Research (CAE-R), given to institutions focused on IA research. CAE-R was added to the program in 2008 to encourage academia development and doctoral research in cybersecurity. CAE/2Y was added in 2010 to involve two-year institutions and other training institutes in cybersecurity education. This link provides the list of institutions state by state that are DHS/NSA certified: https://www.cybersecuritymastersdegree.org/dhs-and-nsa-cae-cd-designated-schools-by-state/.

To understand which schools are considered leaders in cybersecurity education, the Ponemon Institute was directed by HP to conduct a study on the best schools for cybersecurity education in 2014. The report rated schools according to respondents' rankings based on five criteria, namely academic excellence, practical relevance, experience and expertise of program faculty, experience and background of students and alumni, and professional reputation in the cybersecurity community. Respondents were practitioners in information technology (IT) and IT security drawn from several industries across the nation. The researchers gave respondents a list of about four hundred schools in the United States, asking them to pick their five top choices and ranking them according to the five criteria given above. According to the study, the Ponemon Institute reported the following as the twelve top-ranked schools chosen by the respondents:

1. University of Texas, San Antonio
2. Norwich University
3. Mississippi State University
4. Syracuse University
5. Carnegie Mellon University
6. Purdue University
7. University of Southern California
8. University of Pittsburgh
9. George Mason University
10. West Chester University of Pennsylvania
11. U.S. Military Academy, West Point
12. University of Washington

The Ponemon Institute further analyzed these programs according to respondents' choices and found ten common characteristics among the top ten schools:

1. Interdisciplinary program that cuts across different but related fields, especially computer science, engineering, and management
2. Designated by the NSA and DHS as a CAE in information assurance education
3. Curriculum addresses both technical and theoretical issues in cybersecurity
4. Both undergraduate and graduate degree programs offered
5. A diverse student body, offering educational opportunities to women and members of the military
6. Faculty composed of leading practitioners and researchers in the field of cybersecurity and information assurance
7. Hands-on learning environment where students and faculty work together on projects that address real-life cybersecurity threats
8. Emphasis on career and professional advancement
9. Courses on management, information security policy, and other related topics essential to the effective governance of secure information systems
10. Graduates of programs placed in private- and public-sector positions

These points show a differentiation in the quality of education received by cybersecurity experts at these academic institutions and show why the CAE designation is significant in cybersecurity education as well as other criteria that are crucial to the success of any academic institution that wants to be considered a leader in cybersecurity education.

CollegeChoice.net (https://www.collegechoice.net/rankings/best-masters-in-computer-forensics-degrees/) rated the top ten master's programs in computer forensics in 2018. The schools listed were:

1. Carnegie Mellon
2. Johns Hopkins
3. Purdue
4. Boston University
5. George Washington University
6. University of Illinois at Urbana-Champaign
7. Rochester Institute of Technology
8. George Mason University
9. University of South Florida
10. University of Alabama at Birmingham

As you can see from both the Ponemon and College Choice studies, there are some repeat offenders with CMU, Purdue, and Mason on both lists, and all are R-1 rated research universities.

Conclusion

Academic credentials in the field of cybersecurity and digital forensics are a critical part of your overall professional portfolio. Forensic sciences in general require a college education since as a forensics expert you will be required to testify in some court proceeding at some point in your career. How you go about putting your academic portfolio together is unique to you. There is no one way to go about it. For the most part, federal positions in digital forensics require a college degree.

If you have a nontechnical undergraduate degree or are looking for more exposure to digital forensics, then a graduate degree is an option. Brick-and-mortar programs are generally better from a retention perspective, but your own personal situation will determine whether an online program or an in-class program is best.

Cost is always an issue; at least it was an issue for me. Look for a program that offers the best value. In-state tuition or two-year community colleges generally offer the best price and value.

CHAPTER 5

Cybersecurity Career Opportunities in the Field of Criminal Justice

We already discussed the need for cybersecurity professionals and the growing need for digital forensic investigators in previous chapters. In this chapter, we will identify where the opportunities are for criminal justice students within the field of cybersecurity and digital forensics. These jobs may be in the public or private sector. Many opportunities in the public sector exist within state and local law enforcement agencies. However, there are other available opportunities within the federal government. There are also numerous opportunities within the private sector. The most important factors that can make or break a student's professional career are their personal determination and persistence and their ability to adapt to a career field that is unique and dynamic, to foresee where opportunities exist, and to utilize the resources available to them to envision a career for themselves.

The current growth in cybercrimes and cyberattacks in both the private and public sector indicates that there will be an increased demand to prevent and respond to such activities. The rate at which ransomware attacks are increasing is especially alarming. Ransomware attacks, being the fastest-growing cybercrimes, are increasingly affecting many industries, from individuals to local and state governments, as well as the financial and the health-care industries.[1] Dan Lohrmann states, "As escalating cyberthreats continue to grab global headlines, local governments have been hit particularly hard over the past year. From cities to counties to townships, the breadth and depth of attacks have overwhelmed many jurisdictions."[2] Consequently, the State and Local Cybersecurity Improvement Act (2020) was passed to assist states and local governments in their fight against cybercrimes and cyberattacks. Criminal justice students should be taking advantage of opportunities that arise as a result of these circumstances because there will be a need to respond to cybercrimes.[3]

Current Opportunities and Jobs Needing Cybersecurity in Criminal Justice

Many of the opportunities that are within the public sector lie in law enforcement, where interested students can look forward to a career of responding to cybercrimes already committed or preventing future cybercrimes from occurring through the creation of legal and ethical frameworks. These job opportunities are at the federal, state, and local level. In addition to jobs in the government and law enforcement, other jobs exist in the private sector, which may involve consulting depending on what the job entails. There are four important themes that are a part of a risk management plan prepared by the Department of Homeland Security: mitigate, prepare, respond, and recover.[4] This will give students a general idea of the types of jobs that lie in cybersecurity and digital forensics. Students can take advantage of these opportunities if they align with one or more of these four themes in either the public or private sector.

Jobs within the Federal Government (Public Sector)

Cybersecurity jobs fall into the following facets of government: law enforcement, courts, and corrections. The majority of the jobs that are suited to criminal justice students will fall within the response purview (digital forensics) of law enforcement within any of the agencies below. There will also be jobs that are preventative in nature.

Many current job opportunities are within federal agencies like the Federal Bureau of Investigation (FBI), the National Security Agency (NSA), the Central Intelligence Agency (CIA), and the Department of Homeland Security (DHS). Other government agencies that are not necessarily law enforcement in nature will most likely need students skilled in cybersecurity. Organizations like the Department of Justice (DOJ), the Office of Personnel Management (OPM), and other government agencies also need cybersecurity personnel, but those personnel are geared more for prevention. Students need to be creative if they want jobs within the federal, state, or local governments that are not law enforcement in nature. However, most of these jobs are only open to citizens of the United States. Students who are not citizens will most likely have to look for jobs in the private sector.

Jobs within State and Local Governments (Public Sector)

As far back as 2014, the Police Executive Research Forum suggested all police departments respond to cybercrimes.[5] As of 2020, the level of cybercrime activities has continued to rise, and it will be in the best interest of every county to create a cybercrimes/cybersecurity response, investigative, or liaison unit to collaborate with other agencies that can handle cyber-related responses. Ideally, every state should have a coordinated cybersecurity strategy. We believe this will create a need for experts in the field, thereby creating more opportunities for criminal justice students. There are investigative jobs (digital forensics) and preventive jobs (cybersecurity) at the state and county levels.

Most investigative jobs at the state and local levels will be in law enforcement. Many law enforcement jobs will be in the Department of Public Safety (DPS). For example, the Texas Department of Public Safety has a computer investigative unit.[6] Local police departments in urban counties with cities like Houston, New York, Chicago, and Los Angeles have special divisions that respond to cybercrimes. More rural counties defer to larger police departments and the FBI to respond to cybercrimes.

Jobs in state governments will most likely fall within the city, therefore, within local governments or counties. For example, in the state of Texas, there are cybersecurity jobs in the state[7] but some more in large cities like Houston,[8] Austin,[9] and Dallas.[10] Again, students will have to be creative to search for such jobs.

In Houston, Texas, for example, the City of Houston has both an investigative financial unit and a cybercrime unit under the Houston Police Department[11] and hosts its cybersecurity activities under the Houston Information Technology Services (HITS).[12] Other urban jurisdictions in the state have similar institutions that handle prevention through an information technology/security department and investigation through the police department.

Courts and Corrections

There are no clear-cut jobs involving cybersecurity in corrections and courts. However, it is important to note that digital forensic investigators sometimes need to give testimony in courts. To properly provide this testimony, investigators require a certain level of expertise.

An investigator's expertise is needed when a case is taken to court to give "expert testimony." This expert witness may be at a deposition (in an attorney's office) or at trial in a courtroom.[13] In both cases, the investigator needs to be prepared to use their skills, education, and training and utilize standard scientific techniques to prepare evidence that is essential to understanding the facts of a case. An expert witness can be called to testify for the defense as well.

Jobs within the Private Sector

Even though criminal justice students are inclined to look for job opportunities in law enforcement, there are many cybersecurity and digital forensics jobs in the private sector. A student must once again be creative, open-minded, and assertive in envisioning a career path in cybersecurity or digital forensics. They will need to determine their fate by deciding on what education and/or certification skills they need to acquire to land a job. There are various tools that students have at their disposal to determine a career path either in the public or private sector. We will focus on two, the Cyberseek tool created by National Initiative for Cybersecurity Education (NICE) and the National Institute of Standards and Technology (NIST) NICE Workforce Framework.

National Initiative for Cybersecurity Education (NICE) Cyberseek

According to Cyberseek,[14] a website that provides data on the supply and demand of cybersecurity professionals, there are five main career pathways or feeder roles for individuals seeking careers in cybersecurity:

1. networking
2. software development
3. systems engineering
4. financial and risk analysis
5. security intelligence

These five feeder roles are further divided into three levels, entry-level, mid-level, and advanced level. Obviously, not all feeder roles are directly suitable to criminal justice students. Off the bat, financial risk analysis and security intelligence are roles that mirror law enforcement and are therefore best suited to criminal justice students. However, networking and systems engineering are feeder roles used for digital forensics, and students need those skills to succeed. Having said that, some of these roles may involve less technical skills than others. This does not mean a criminal justice or non-technical-oriented student cannot seek roles in areas with more technical skill sets. We believe they can transition by getting further education, training, and certifications.

Entry-level jobs in networking career paths are usually listed as cyber-security/specialist technician, cybercrime analyst/cybercrime investigator, incident analyst/responder, and IT auditor. A student who is interested in the private sector would be best suited for the cybercrime analyst/investigator and incident analyst/responder entry-level roles. However, we suggest that students examine all roles listed by Cyberseek.[15] We have provided a full list of those roles in appendix A. Appendix B is information extracted from appendix A, a shorter version showing job roles best suited to criminal justice students.

National Institute of Standards and Technology (NIST) Workforce Framework[16]

Another tool important in determining roles in cybersecurity and digital forensics is the National Institute of Standards and Technology (NIST) Special Publication 800-181 titled "National Initiative for Cybersecurity Education (NICE) Cybersecurity Workforce Framework." This framework was jointly created by NIST and NICE to restructure cybersecurity roles and streamline job positions to each area of specialty. The framework divides the cybersecurity workforce into seven main specialty areas each with its own subsets. The framework further describes knowledge, skills, and abilities (KSAs), work/role definitions and descriptions, as well as the responsibilities that fall under those roles.

Below is a description of each of these seven specialty areas:

Table 5.1.   NICE Framework Workforce Categories

Categories	Descriptions
1. Securely Provision (SP)	Conceptualizes, designs, procures, and/or builds secure information technology (IT) systems, with responsibility for aspects of system and/or network development.
2. Operate and Maintain (OM)	Provides the support, administration, and maintenance necessary to ensure effective and efficient information technology (IT) system performance and security.
3. Oversee and Govern (OV)	Provides leadership, management, direction, or development and advocacy so the organization may effectively conduct cybersecurity work.
4. Protect and Defend (PR)	Identifies, analyzes, and mitigates threats to internal information technology (IT) systems and/or networks.
5. Analyze (AN)	Performs highly specialized review and evaluation of incoming cybersecurity information to determine its usefulness for intelligence.
6. Collect and Operate (CO)	Provides specialized denial and deception operations and collection of cybersecurity information that may be used to develop intelligence.
7. Investigate (IN)	Investigates cybersecurity events or crimes related to information technology (IT) systems, networks, and digital evidence.

Note: Available at https://nvlpubs.nist.gov/nistpubs/SpecialPublications/NIST.SP.800 -181.pdf, page 11.

From the description of the categories above, it is clear that the Investigate (IN) area fits criminal justice students. Some of the specialty areas like Securely Provision (SP), Collect and Operate (CO), and Operate and

Maintain (OM) are for highly skilled and technical fields like security software engineers, programmers, and architects. However, others like Protect and Defend (PR) and Analyze (AN) have some roles suited to criminal justice students. Moreover, we see criminal justice and other nontechnical students fitting in many of the Oversee and Govern (OG) roles as they advance in their cybersecurity careers. The NIST SP 800-181 document also clearly identifies the roles, tasks, and knowledge skills needed to become successful cybersecurity professionals. The full table including the job roles; knowledge, skills, and abilities (KSAs); as well as the tasks required for each of the specialty areas can be found on the website at https://nvlpubs.nist.gov/nistpubs/SpecialPublications/NIST.SP.800-181.pdf. It is too detailed to present here. However, appendix C has an abbreviated version that gives students an idea of what some of these roles entail.

One thing that should be pointed out is that appendix C is listing work roles, not job titles per se. What this means is that a particular position could include a number of these work roles depending on the size of the organization as well as the division of labor. For example, the threat warning and exploitation roles could be handled by the same analyst/engineer.

Other Important Roles

There are many other roles that may not be included in the Cyberseek or NIST framework because they may fall under other responses to cyber-crimes or specific areas of cybersecurity or digital forensics not included in these tools or frameworks. Some of these roles may have similar tasks or KSAs as identified roles above. A list is provided for any interested student to search and figure out if they are suited for such roles. Here is a categorized list, which is by no means exhaustive:

Mitigate and Prepare (Geared toward cybersecurity):
- Information Assurance Auditor
- Ethical Hacker
- Cyber Threat Manager
- Intel Analyst

Response (Geared toward digital forensics)

- Incident Responder
- Cyber Investigator
- Digital Forensics Examiner
- Digital Forensics Analyst
- Computer Forensics Analyst
- Internet Crimes Against Children (ICAC) Investigator

Recover (Geared toward digital forensics)

- Digital Forensics Examiner
- Cyber Investigative Technician
- Cyber Threat Manager
- Internet Crimes Against Children (ICAC) Investigator

CHAPTER 6

Planning Your Path into the Cybersecurity and Digital Forensics Field

The cybersecurity career path is not as straightforward as we have previously alluded to for most people, let alone criminal justice students. This path evolved as a result of the challenge of responding to the criminal use of technology, cybercrime, and cyberattacks. Many companies and governments started to respond to these breaches in an ad hoc manner until it became obvious that a special skill set was needed to address these breaches.[1] In fact, in an article titled "The Cybersecurity Profession Has a Clear Career Path. LOL. Just Kidding," Candy Alexander calls the successful progression in various professionals' cybersecurity careers "dumb luck" and the creation of a career path "reactive."[2] The Department of Homeland Security (DHS) and the National Institute of Standards and Technology (NIST) created the National Initiative for Cybersecurity Education (NICE) and National Initiative for Cybersecurity Career and Studies (NICSS), before which there was no clearly marked career pathway in cybersecurity. Currently, the vast number of career titles, roles, tasks, knowledges, skills, and abilities make it difficult for the average person to map out or navigate a clear career path. This chapter will attempt to provide a guide for students to do just that. The guidance given here refers to cybersecurity in general, but this road map may also be tailored to a student interested in digital forensics as well. For specific information on digital forensics, please refer to chapters 3, 4, and 7.

Since students may require assistance navigating a career, they may need the aid of specific people in the cybersecurity community. Therefore, the information in this chapter is for students as well as every member of their university and the local community that make decisions that impact the careers of students. Members of this community should include top school administrators (president, provost), career guidance counselors, faculty, mentors (CEOs and top officials of organizations and companies in the immediate environment of the school), city/local government officials (mayor, city council), other local school administrators (head of K–12 school districts), and officials in neighboring schools (like community colleges).

Before a student can map a career, there are certain questions they must answer to determine if they are making the right choice:

- The student will need to determine their career goal, or at the least have a rough idea of what they would like to do within this field, a general cybersecurity role or specific subcategory like digital forensics. In other words, what aspect would they like to pursue? Do they see themselves in law enforcement? If so, they may want to pursue digital forensics (see chapters 3, 4, and 7). Do they want to be an analyst, a consultant, or at the ethical or legal end of cybersecurity? If they want to go to the private sector, they can still have a career in digital forensics (see chapter 5).
- The student will need to determine their current skills and/or what additional skills they need to attain that career they are targeting (see chapters 3, 4, and 5).
- The student will need to determine if they need education (traditional), training, or both. Another option they might consider is if the certification route would get them the needed skills (see chapters 4 and 5).
- The student will need to determine which resources are required to attain their desired career. How much would they need to get the education, training, and/or certification? Where will they get the resources to pay for that education? Students who don't have the resources to pay for a formal education may try the training route, which is shorter and may or may not be less expensive than getting a bachelor's degree (see chapter 4).
- Finally, the student will need to envision where they see themselves in the future. Are they aiming for a career in cybersecurity or digital forensics within a law enforcement agency, within a government organization, or within the private sector? In other words, they should envision what that career would look like in five to ten years. For example, are they aiming for a top management position?

If a student is able to answer these questions, it becomes easier to map a career. Obviously, these are not simple questions that can be answered in a short amount of time. The student will have to be creative, ingenious, resourceful, and persistent. A road map is necessary for making these decisions. Therefore, we provide a rough guide and information to aid students in making those decisions.

In chapter 5, we explained how the NICE framework and the Cyberseek tool provide some guidance for career roles in cybersecurity and digital forensics. The NICE framework provides a standardized career-guidance tool based on job functions, roles, tasks, and knowledge, skills, and abilities, while the Cyberseek tool is an interactive guide that shows progression in cybersecurity roles and how an individual can progress in their career. Cyberseek also shows where jobs exist in cybersecurity as well as suggested training, education, and certifications needed for each job. These roles include digital and computer forensics among many other suggested roles. Students can start their career search by looking at the roles and tasks and determining where they see themselves in five to ten years. In this chapter, we will discuss career mapping using cybersecurity first; then we will be specific, by choosing digital forensics as a specific career choice, which is usually the most suitable career path for a criminal justice student.

A Proposed Model for a Successful Cybersecurity Education and Career

EDUCATION

As of 2018, a university degree was no longer required for some tech jobs. Both an article by Courtney Connley for *CNBC Careers* and Glassdoor listed IBM, Google, and Apple as tech companies that no longer require a degree.[3] They require a prospective employee to have hands-on experience or to have completed a boot camp. For candidates applying for a software engineer position, they must complete a coding challenge to be considered for employment. While many criminal justice students may not have the proper skill set to complete this challenge, many law enforcement officers who become digital forensic experts in their jurisdictions had no prior coding or technical skills. They simply learned on the job and got trained. It is difficult to predict if this trend will continue.

While training (like certifications) gives students the skill sets for the job, education is a well-rounded approach that prepares a student for the real world, not only as a skilled employee in their respective field but as a person who can solve problems. That is not to say only students who go to a university or college are problem solvers. However, a university or college campus is meant to give students more than just technical skills; it also enables them with problem-solving and critical-thinking skills. We recommend a degree for anyone seeking jobs in the federal government. But there are many people who don't need to go to college to develop problem-solving skills or be critical thinkers. Therefore, it is left to a student to determine which route they would like to take to build a career in the field of cybersecurity or digital forensics. If a student wishes to choose the education route, at least a bachelor's degree is recommended for entry into the field.

A student can get an associate's degree, a bachelor's degree, or a master's degree in many cybersecurity-related fields depending on their answers to the five questions above. However, just as discussed at the beginning of the chapter, due to the vast number of roles, tasks, and knowledge, skills, and abilities (KSAs), it can be a daunting path to navigate without a guide. Because the number of schools offering cybersecurity and digital forensics degrees are not as many as other fields like criminal justice or business, students can find it very confusing to understand what career roles are mapped to certain education or training. The roadmap in figure 6.1 below explains that process for a student. This roadmap guides students on how to navigate a career in cybersecurity and digital forensics.

Students can find opportunities around them to receive an associate's degree. For a general guide, we will refer students to the Center of Academic Excellence (CAE) designated schools list—with 2Y signifying a two-year college degree—found at National Information Assurance and Education Training program's (NIETP's) website (https://www.iad.gov/NIETP/reports/cae_designated_institutions.cfm). The schools are listed by state, which makes it easier to search for a school. Again, not only CAE-designated schools are available for education purposes. There are other schools not designated as CAE that offer degrees in cybersecurity. The student will simply have to do their research to determine what is best for them. The student can follow this same process for information about a bachelor's degree as well. For more information about CAE-designated schools, see chapter 2.

Figure 6.1.   Road map to cybersecurity/digital forensics education/ career. *Figure created by authors*

For a graduate degree, a student should be aiming for mid-level management positions and should follow the same process as searching for a bachelor's. However, they will need to dig deeper to determine what area they would like a graduate degree in. At this point they already have a bachelor's degree and should consider if they would like to switch their area of study or continue in the same direction as their bachelor's. If they are switching their area of study, they will need to determine their skill set and what prerequisites are needed to get into a graduate program. For example, a criminal justice graduate can get a master's degree in cybersecurity in a nontechnical area (see figure 6.2 below). Again, a search of the NIETP website will guide students. A query on search engines can also direct students to such programs.

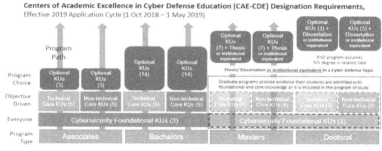

Centers of Academic Excellence in Cyber Defense Education (CAE-CDE) Designation Requirements, Effective 2019 Application Cycle (1 Oct 2018 – 1 May 2019)

Knowledge Units (KUs):

Foundational: Cybersecurity Foundations, Cybersecurity Principles, and IT Systems Components

Technical Core: Basic Scripting and Programming; Basic Networking; Network Defense; Basic Cryptography; Operating Systems Concepts

Nontechnical Core: Cyber Threats; Policy, Legal, Ethics, and Compliance; Security Program Management; Security Risk Analysis; Cybersecurity Planning and Management

Figure 6.2.   CAE designation requirements for cyber defense for cycle 2018-2019. *Adapted from National IA Educational and Training Programs, "NSA/DHS National CAE in Cyber Defense Designated Institutions"*

TRAINING AND CERTIFICATIONS

If students decide to take the training route, they need to search for the best training institutes. First, they must decide a career field of choice or skill set. If a student is interested in criminal justice, law enforcement, or digital forensics, they would need to identify trainings and certifications in that area. For digital forensics, a student can investigate certain companies and what digital forensic certifications are available (see chapter 4).

Some universities partner with specific companies to offer certificate courses in some specialty areas. Coursera is an example of such a company. Other companies like Cybrary partner with vendors to provide short-term training and courses. While these companies have a few free training courses available, registration is required, and payment of a subscription fee is needed to access other advanced courses. Further instruction in any course will require the payment of the subscription fee for Cybrary, while Coursera needs the participant to pay for the certificate for some of its free courses. There are other companies that offer certifications in digital or computer forensics. Here are some examples:

1. Infosec: https://resources.infosecinstitute.com/category/computerforensics/introduction/computer-forensics-certifications/#gref
2. SANS: https://digital-forensics.sans.org/training/courses and https://digital-forensics.sans.org/certification/gcfe

3. International Association of Computer Investigative Specialists: https://www.iacis.com

A certification like SANS GIAC is not for beginners but requires an individual to be employed in a forensics capacity or with law enforcement.

Other companies offering cybersecurity certifications for students who want to go that route include:

1. CompTIA Security: https://www.comptia.org/certifications
2. (ISC)²: https://www.isc2.org/Certifications
3. ISACA: https://www.isaca.org/training-and-events/cybersecurity
4. Infosec: https://www.infosecinstitute.com

Prospective students can also search the internet or the Cyberseek tool (see chapter 5) for more certifications.

Other Activities That Are Important for Students' Success

In addition to a student having a career plan, there are other activities that can help the student's career. Such activities include networking with hiring agencies and professionals in the field, attending conferences, and getting involved in campus activities that will expose them to jobs in the field.

NETWORKING

According to Jennifer Bonds-Raacke, John Raacke, and Samantha Elliott, it is important for students to network with professional peers. They state,

> Yes. You should be networking . . . now. We know it can be daunting to think about building a professional network as an undergraduate or even graduate student. But the reality of the situation is that building a professional network will likely open many doors and possibilities for you. And most importantly, the sooner you start to network, the sooner you can start gaining additional knowledge and skills to help you do your job better or get that job you want.[4]

While the writers belong to the professional field of psychology, the truth can be said of any profession. Career professional sites like LinkedIn,

ZipRecruiter, and the likes were created on the basis of networking relationships and creating opportunities for job seekers. Other career-focused websites like Indeed and social networking sites like Twitter and Facebook are also great for networking.

CONFERENCES

There are many conferences where students can learn and be exposed to cybersecurity and digital forensic concepts and issues. While many of these conferences are not created for students, they are exceptional tools for learning, gaining some exposure, networking, and job hunting. Some schools sponsor students so they are able to attend such events just as they do for other academic conferences. Students can attend as a group with faculty or industry mentors. Attending conferences should also be a part of their future professional life and could be the beginning foundation of many rewarding relationships.

Some of those conferences are:

1. National Initiative for Cybersecurity Education (NICE) Expo conferences
2. Digital Forensics Research and Workshop (DFRWS) Conference
3. SANS conferences (regional conferences on many topics)
4. International Conference on Digital Security and Digital Forensics (ICDSDF—USA)
5. Techno Security and Digital Conference
6. Internal Conference on Criminal Procedures and Digital Forensics (ICCPDF)
7. International Conference on Digital Forensics and Evidence (ICDFE)

School Career Advancement Activities

INTERNSHIPS

Other activities that can help students get their foot in the door include internships and apprenticeships. In almost every career field, internships are a great way for students to learn, work, and discover if their respective field is one they want to be in. An internship is "the position of a student or trainee who works in an organization, sometimes without pay, to gain work experience or satisfy requirements for a qualification."[5]

Although there are several students that attain internships in their junior year, one should aim to acquire an internship as soon as possible. Some students find out during their internships that the career choice they envisioned was not actually what they wanted, while many others absolutely love their career choice. An internship does two things: On the side of the student, it helps them narrow their career choices and decide on what to do, as well as helps with employment opportunities. For an employer, it aids in finding potential employees that can be seamlessly inserted into the company dynamic.

APPRENTICESHIPS

NICE, in realizing the gap between the supply and demand of cybersecurity professionals, started to promote apprenticeships to enable the transfer of the needed skills to the new generation of young workers. Apprenticeships were promoted to bridge the gap between two generations in a field where there are constant changes in technology. An apprenticeship is "an industry-driven, high-quality career pathway where employers can develop and prepare their future workforce, and individuals can obtain paid work experience, classroom instruction, and a portable, nationally-recognized credential."[6] Apprenticeships are a great way for students to learn on-the-job skills while also being compensated. The difference between apprenticeships and internships is an internship is a temporary arrangement for an extended period of time, while an apprenticeship is ongoing through the student's education. In most cases during an apprenticeship, the employer pays the tuition of the student and the student works in the company/organization while still going to school. In other cases, companies receive a grant that they use to sponsor students. The advantage of an apprenticeship is that it provides the transfer of knowledge and on-the-job experience, which enables hands-on learning, as well as the funding of educational opportunities.

CLUBS AND SOCIAL ORGANIZATIONS

Clubs and social organizations on campus are another way for students to form relationships, network within and outside the college community, and gain knowledge on subjects they would not typically learn in a classroom. Clubs and social organizations also provide students with the united goal to raise awareness, raise money, and push for changes that cannot be done by one person. Once a club or campus organization has been formed, it can be used by students to push agendas to the school and immediate environment or community.

COMPETITIONS

Competitions allow students to join their colleagues in the field to enhance their skills. The Department of Homeland Security (DHS) has provided a platform for various competitions through its website.[7] Other organizations also offer cybersecurity competitions, and there are a few digital forensics competitions. The purpose of these competitions is to give participants a way of meeting their peers, sharpening their skills, and networking with other like-minded people.

The Role of Colleges and Their Community

Every college has a role in providing a nurturing environment for the growth of a thriving cybersecurity community. The college or university needs to be the catalyst in providing students with opportunities that will enable them to grow as resourceful members of the field. It is imperative that the college act as an incubator, enabling all the pieces to work together. The college should ensure that it places itself as a member of the general cybersecurity and law enforcement community by doing the following:

- Creating a center/hub for cybersecurity and digital forensics activities centered on the students' success. This hub is necessary to coordinate activities across the campus to ensure that resources are available and utilized properly, not just for technologically oriented fields. In fact, it is important that a school has a top-driven approach to cybersecurity education, meaning the university's management has to have the belief in and support for cybersecurity education.[8] It will be the duty of the university to ensure it places itself in a position to provide a well-rounded education for its students, obtain resources that students will need to be successful in the field, engage the immediate community to hire its students, and engage the community in all other ways that will lead to the success of its students.

- This center/hub should connect students to internships, apprenticeships, community-engaging activities, competitions, mentorship, etc.
- The center should also be looking at making a school a Center of Academic Excellence (CAE) as discussed in chapter 2.

The immediate community also plays a role in a student's success by partnering with other schools, colleges, universities, and K–12 schools. This community should include law enforcement, local cybersecurity and digital forensics companies, local/city government officials, and college administrators. The members of the community can ensure student engagement, guaranteeing that there is a continuous flow of activities and students to the colleges, which will then lead to the continuation of success in the community.

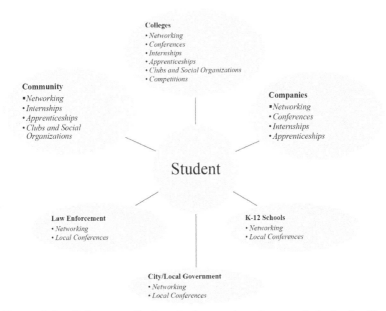

Figure 6.3.   Cybersecurity incubator centered around students. *Figure created by authors*

Getting the Job and Entering the Digital Forensics Field

There are three factors that affect entering any field, and digital forensics is no exception:

- education
- experience
- persistence not annoyance

We've talked about the education part in detail, but I want to address experience and persistence. So how do you get experience when first starting

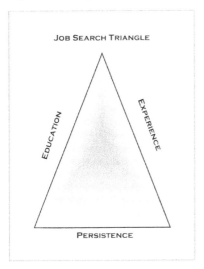

JOB SEARCH TRIANGLE

EDUCATION

EXPERIENCE

PERSISTENCE

Figure 7.1. Job search triangle.
Figure created by authors

out? Digital forensics is one of those fields where it's possible to gain some experience without actually working in the field. I'm not talking about internships here although internships are a great way to get some experience.

A motivated individual can actually get some valid experience with a minimal outlay of capital resources since you can build your own digital forensics lab and practice. "Wait a second," you say. "It's too expensive to build my own lab." Well, actually it's not. A couple of laptops, a few virtual machines (VMs), and some open-source/free tools can start you down the path quite well.

Setting Up a Home Digital Forensics Lab

First, you will need a couple of computers, probably laptops, but any computers will do. One computer will run Windows and the other Linux. What about Apple Macs? If you have one, great. You can definitely include it in your lab inventory, but Macs are usually expensive, and Apple does not like people running Mac OS as a virtual machine client although it is possible. I have never run Mac in a VM and will not cover it here, but a motivated person can run a Google search and figure it out, though you will likely be violating Apple's terms of service.

You should also have a five-port ethernet switch ($20), a USB DVD drive ($30), and an external 2 TB drive ($60) and some assorted ethernet and USB cables. "Wait a second, what do you mean ethernet cables? Are you in the dark ages? I use wireless." Wireless is fine for many things, but your digital forensics lab needs to stand on its own as a closed or at least semiclosed system. Also, you will need a few SD cards and thumb drives to practice.

Your second computer can easily be a Raspberry Pi. A CanaKit Raspberry Pi 4 4GB Starter MAX Kit, 64GB Edition runs about $115 and is a full-blown distribution (aka distro) of Linux. An old working laptop that someone is getting rid of will also work well. Linux is a much lighter OS than Windows so it will run perfectly well on older equipment.

For your Windows computer, the more robust the better, but at a bare minimum the machine should have 16 GB memory, an I7 processor, and a 500 GB drive. More memory, an I9 or Itanium processor, and a 1 TB SSD or larger is always better but more expensive. You can probably find a new base laptop for $600 to $700. For less than $1,000, you have the hardware for your lab.

Let's look at available software that is essentially free:

AUTOPSY

Autopsy (https://www.sleuthkit.org/autopsy/) is a free, fully functioning digital forensics platform. It will run on Windows, Linux, and Mac. As listed on its website, some of its features include:

- Multi-User Cases—More than one person can work on the same case.
- Timeline Analysis—Has a functional graphic that allows you to plot file system time-based events.
- Keyword Searching—Looking for Bob or Lucy may not be very productive (too many false positives), but you can search your media for any text string (the longer the string the better).
- Web Artifacts—What websites did the user go to? The Registry is the master configuration database for Windows. Autopsy lets you view the Registry for the most recently used (MRU) resource or if a thumb drive was used.
- LNK File Analysis—Desktop shortcuts can be examined.
- Email Analysis—Examine emails sent and received by the user.
- EXIF, which mean Exchangeable Image File Format—Allows you to look at photographs and find out where the photo was taken and by what device.
- File Sorting—Organizing files to make it easier to review.
- Media[1]—View videos and photos with a Thumbnail viewer.
- Read the following File Systems—NTFS, FAT12/FAT16/FAT32/ExFAT, HFS+, CD-ROM, Ext2/3/4, Yaffs2, and UFS from The Sleuth Kit.[2]
- Filter using hash sets.[3]
- Bookmark items that are evidentiary.
- Add comments to better explain an artifact.
- Extracts Unicode.[4]
- Signature based file type detection.[5]
- Support Android devices.[6]

- Input Formats—raw/dd or E01.[7]
- Reports can be formatted in HTML or XLS.

As you can see, it is fairly versatile for a free tool. There are law enforcement agencies that use Autopsy.

FTK IMAGER

AccessData (AD) has provided FTK Imager (https://accessdata.com/product-download/ftk-imager-version-4-2-0) for free for many years. You do need to register with AD. FTK Imager is an extremely capable imaging tool. It can image just about all media to a varied list of image formats. You can collect bit-for-bit images, logical images, memory, and Windows Registry files. FTK Imager also has basic viewing capability. Make sure that you get FTK Imager from AccessData and not some no-name site since the no-name site could be providing a weaponized version of FTK Imager. This is true for any tool that you download from the internet. AccessData was purchased by Extero in December 2020.

PALADIN

Paladin (https://sumuri.com/software/paladin/), from Sumuri, is a Linux-based, full-featured digital forensic platform that incorporates Autopsy. Sumuri asks that you pay $25 for Paladin. It is well worth the cost. You can always cheap out and download it for free.

Using Autopsy, FTK Imager, and Paladin, you have digital media covered. Mobile device forensics is a little more problematic with free tools, but there are some options. Autopsy has some Android capability. Jessica Hyde, director of digital forensics, Magnet Forensics, one of the best mobile device examiners on the planet, and a George Mason University Digital Forensics graduate and adjunct professor, advised that for IOS imaging you can use Magnet Acquire (freeware) or use iTunes to obtain possible backups. If the device is jailbroken[8] or if you can jail break it, you can follow the directions in the below posts for a full File System. Magnet Acquire will also get you a full file system on a jailbroken device including the new checkra1n jail break easily: https://www.mac4n6.com/

blog/2018/1/7/ios-imaging-on-the-cheap-part-deux-for-ios-10-11 https://www.4n6files.com/2019/05/creating-file-system-image-of-ios12.html (this one is a result of a project by a student in Jessica's Mason class).

For analysis, Jack Farley has made a tool for reading iOS backups: https://github.com/jfarley248.

For full file system opensource analysis, you can use iLEAPP from Alexis Brignoni: https://github.com/abrignoni/iLEAPP. There is a cool video of it working here: https://www.youtube.com/watch?v=yr7ohaJzW5A.

You can also try Apollo from Sarah Edwards: https://github.com/mac4n6/APOLLO.

Looking for the Job Posting

When talking about law enforcement careers in the United States, we need to separate federal, state, and local/county jobs. At the federal level, job opportunities are generally competitive and are posted on USAJobs (https://www.usajobs.gov/).

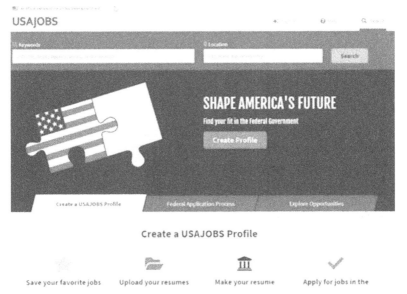

Figure 7.2.   USAJobs. *USAJobs*

USAJobs is a main portal for federal jobs, but it's not always the only portal. Some federal agencies do have their own sites. USAJobs allows you create an account and set up a profile where you can upload resumes, save job searches, and link to job postings. You can even save your searches and have USAJobs email you when new postings become available. Last, but definitely not least, you can apply through the USAJobs portal.

One particular hiring situation to look for is whether the position is a direct hire. A direct hire is where a determination is made that there is a critical need for a particular job or jobs.[9] What this means for the job seeker is that the process can be more streamlined. Public notice must be given by the hiring agency, but the agency can make a hiring decision without having to competitively rank, consider veterans preference, or limit hiring to the top three candidates (rule of three). Usually when a position is a direct hire, there may be a name in the posting where the applicant can directly interact. Similar to a direct hire, agencies like the FBI may maintain a pool of qualified candidates. Once qualifying for the pool, a selection can be directly made.

Here is an example of a link for a GS13-14 HSI (Homeland Security Investigations[10]) IT (Information Technology) Cybersecurity Specialist Vacancy Announcement: https://www.usajobs.gov/GetJob/ViewDetails/560803600. This position is within HSI and will be doing the investigative/proactive cyber mission and not a traditional IT type "protect the network" position. A start date is negotiable in the event someone still has another semester. If anyone is going to put in, they should highlight their relevant experience.

So you've found a posting on USAJobs, and you are excited to apply. Is the job in the geographical area that you want to work? Do you meet the basic qualifications for the grade of the job that is being posted? General Schedule (GS) pay scale grades have minimum requirements for each grade. Grades GS-2 through 4 are usually entry-level jobs that do not require a college degree. There are usually no digital forensics positions at these grades. GS-5/7 grades require a bachelor's degree, and you can qualify for the GS-7 if you have a GPA of 3.0 or better. GS-9 requires a master's degree and GS-11 requires a PhD. The grades we're targeting for entry-level digital forensics examiners are GS-5/7/9.[11] Grades GS-10 thru 15 usually require that you meet the requirements for the lower grade, then have additional specialized experience for that job.

Before you hit the apply button, you need to take a close look at the job requirements usually, but not always, called knowledge, skills, and abilities (KSA). For each KSA, you need to directly map your education and experience. Take each KSA or requirement and write a specific response. The table below is an example of how you can evaluate your skills by KSA.

Table 7.1.   KSAs and Your Skills

KSA	Your Skills
Do you meet the minimum qualifications for the position grade?	Yes/No If no, you do not need to proceed.
Computer systems and digital evidence including computer hardware, processes, operating systems, application software, utility programs, storage, electronic mail systems, intrusion tools, and Mac, Windows, and Linux operating systems	How you meet this KSA
Digital forensic processes and procedures	How you meet this KSA
Digital forensics tools and certifications (e.g., FTK or Truxton Certifications, SANS/GIAC, IACIS Forensic Examiner, Cellebrite, or XRY Certifications)	How you meet this KSA

The number of KSAs will vary, but generally expect three to five individual KSAs. "Why do I need to go through this exercise? My resume/CV covers all the great things that I can do." My experience as federal government application reviewer is that applicants that do not effectively answer the KSA questions may not get to a human reviewer. Given the volume of applicants for federal positions, you want your application to stand out. Effectively answering the KSA questions can get you to the next level. Not answering KSAs will most likely pull you out of contention.

The National Institute of Standards and Technology (NIST) has put together an extensive list of KSAs for the federal cyber positions in NIST Special Publication 800-181 "National Initiative for Cybersecurity Education (NICE), Cybersecurity Workforce Framework" (https://nvlpubs. nist.gov/nistpubs/SpecialPublications/NIST.SP.800-181.pdf). The list in appendix A will provide you with comprehensive lists on knowledge, skills, and abilities broken out. There are literally more than a thousand KSAs, but don't get nervous. You don't have to be an expert in everything. NIST put this publication together to help employers, but it can also help the ones that are looking for work if you know what's expected ahead of time.

State and local digital forensic position postings vary wildly. Identify the agencies for which you are interested in working. Find out how the agency posts for open positions. If you can, register with the job portal the agency uses. I know that it's old school, but a phone call can't hurt after you've done a little online research. Sometimes websites either tell you too much or not enough. If you're lucky enough to get a living, breathing body on the other side of the phone, you can ask some questions like:

- What's the process for hiring digital forensics examiners?
- Are your examiners sworn officers, nonsworn, or a mix?
- Are you looking for any specific degrees or certifications?
- Are there any physical performance requirements to the position?
- Would it be possible to talk to someone in the lab?

Yes, these are basic questions, but they are also questions that can give you insight into what the agency is looking for. See if the agency sponsors a career day. If so, attend the career day and locate the people that are involved in digital forensics examiner positions.

Persistent but Not Annoying

Obtaining any law enforcement position requires a certain amount of persistence. Job postings may be inconsistent, meaning only when funding is available and/or the funded staffing level is approved. It may not be a bad idea to contact your local law enforcement and ask them how they hire digital forensics examiners. Be on the lookout for any presentations or talks being given by law enforcement agencies. Even if the talk is not digital forensics specific, you can ask the question "How do I go about getting a digital forensics job at any county police department?" Stay in touch with your university's career services unit.

When interacting with people, persistence is usually admired, but there is a fine line between persistence and annoyance.

Personal Story of Persistence

I had wanted to become an FBI Agent ever since I was fifteen years old. My grandfather knew someone at the FBI, and he brought home brochures on the FBI. I was hooked. When I graduated from college, I applied to the FBI as soon as I was eligible. I got rejected. I contacted the FBI, asked why I was rejected, and was advised that I needed to file a Freedom of Information Act (FOIA) request. I filed an FOIA request. I got my local congressman, who I did not know, involved, and the FBI reopened my case, and I got hired.

Without persistence, I never would have gotten the job.

At no time in the process of getting hired by the FBI did I badger anyone, complain, or otherwise be a pain in the butt. I asked questions, thanked everyone for their help, and proceeded forward. I did not invent this process but listened to people who mentored me on how to go about solving this challenge.

The Resume

Your resume may not get you the job, but it can prevent you from getting the job. Also, make sure that your personal contact network has your resume. Your personal contact network are people you know that may be helpful in getting you a job. Should you have more than one resume? What resume format best suits you? What resume format does your future employer expect to see? Should you have a LinkedIn profile? Is your Facebook outward-facing profile professional?

Art Ehuan, a former FBI Cyber Agent and vice president of Crypsis (now PaloAlto), a digital forensics, incident response, and cybersecurity company, advises that a resume should not be more than two pages. No one is going to read it if it's any longer than two pages. Your resume has to make you stand out and answer the question "Why do I want to hire this person?"

There are basically two types of resumes, a functional resume and a timeline-based employment resume. You should probably have both styles of resume in your inventory to quickly adapt to what the employer wants to see.

Functional Resume Example

Sy Bert Guice
s_guice69@gmil.edu 703-555-1212

I am motivated and trained in the areas of digital forensics and cyber analysis. I am looking for an entry-level position in digital forensics, incident response, pen testing, and cybersecurity.

Professional Skills

- Designed and built my own digital forensics lab
- Supported senior developers and engineers in development of cyber response and analysis tools as a summer intern
- Drive imaging using dd and FTKImager
- Forensics tools include:

  - Autopsy
  - Paladin
  - Wireshark
  - Truxton
  - Magnet Acquire

- Conduct malware analysis using:

  - PE Studio
  - Ghidra
  - PEView

- Scripting experience using:

  - Bash
  - Python
  - PowerShell

- Operating system experience at the command line level:

  - Windows
  - Linux
  - MacOS

Education

BS Cyber Security Engineering
George Mason University
GPA: 3.75

Experience

Summer 20XX Intern
Probity-Truxton
Herndon, Virginia

Fall and Spring 20XW–20XX
Lab Assistant
George Mason University

September 20XV–August 20VX
Utility Team Member
McDougals
Fairfax, Virginia

A functional resume stresses your skills and not necessarily the organizations you worked for. If you are just entering the workforce, then the functional resume may put you in a better light. The timeline-based employment resume also lists your skills, but it does so based on the job where you developed/utilized that skill.

Both resume styles have their strengths and weaknesses. That's why you want to have both resume formats available. This does not mean that you send both formats to a prospective employer. Pick one based on the best intelligence available.

You get a call or an email stating that the State Police would like to interview you for an entry-level digital forensics position. That's great. You're in the door. Now, what are you going to do to seal the deal? Many government agency interviews are prescripted, meaning that they ask the same question to every candidate. This is usually done for legal reasons. Be ready. Do some online research or ask the person who contacted you if there are any resources available to help. Have your elevator speech prepared. What is an elevator speech? You are in an elevator, and the door opens and the Chief of the Department steps in, looks at you, and asks who you are and what you do. You have thirty seconds to a minute to impress. That's your elevator speech. Be able to sell yourself quickly.

Dress appropriately for the interview. For government jobs, business attire (jacket and tie/business dress) never hurts. Bring extra copies of your resume because someone won't have it. Look everyone in the eye but don't leer, shake everyone's hand but don't get into an arm-wrestling

Timeline Resume Example

Sy Bert Guice
s_guice69@gmil.edu 703-555-1212

I am motivated and trained in the areas of digital forensics and cyber analysis. I am looking for an entry level position in digital forensics, incident response, pen testing, and cyber security.

Education

BS Cyber Security Engineering May 20XY
George Mason University
GPA: 3.75

Employment

Summer 20XX Intern
Probity-Truxton
Herndon, Virginia

- Supported senior developers and engineers in development of cyber response and analysis tools as a summer intern

Fall and Spring 20XW–20XX
Lab Assistant
George Mason University

- Conduct malware analysis using:
  - PE Studio
  - Ghidra
  - PEView

September 20XV–August 20VX
Utility Team Member
McDougals
Fairfax, Virginia

- Operated the drive-thru line
- Closed down store
- Prepared food

Developed Skills

- Designed and built my own digital forensics lab
- Drive imaging using dd and FTKImager
- Forensics tools include:

    ○ Autopsy
    ○ Paladin
    ○ Wireshark
    ○ Truxton
    ○ Magnet Acquire

- Scripting experience using:

    ○ Bash
    ○ Python
    ○ PowerShell

- Operating system experience at the command line level:

    ○ Windows
    ○ Linux
    ○ MacOS

contest, smile, and be ready to give your elevator speech and answer questions put to you.

We all know that after reading this book and putting all this great information into practice you will never be turned down for a job, but let's just make the absurd assumption that you had a bad day and didn't get hired. Ask for feedback. What were the reasons why you weren't selected? You may not get answers or, if you are dealing with the federal government, you may have to initiate a Freedom of Information Act request. We all hate rejection, but use this time to critically evaluate yourself so you are better prepared for the next job opportunity.

Take stock of the reasons and be critically objective about yourself. Did you just not meet the qualifications that they were looking for? Did you meet the qualifications but were not able to sell yourself properly? Be honest with yourself, adapt, and get ready for the next interview.

Conclusion

It takes education, skill, and persistence to land the right job. Prior planning prevents piss poor performance. OK, that's enough of the clichés, but you get the idea. Have a plan, and although I hope you get the first digital forensics job you apply for, be prepared for some bumps in the road along the way.

- Do your homework on the agency by which you wish to be employed.
- Have your resume polished and ready to go.
- Work your personal network.
- Get access to job hiring portals (if they exist).
- Find out and attend job fairs with resumes in hand and properly dressed.
- Have your elevator speech ready to go.

Career Advancement in Cybersecurity and Digital Forensics

Recap of Previous Chapters

Previous chapters laid the groundwork for how a student interested in a career in cybersecurity and digital forensics can navigate the field. In chapter 1, we introduced the student to the interesting and diverse field of cybersecurity and digital forensics. Chapter 1 also describes what cybersecurity is and what the cybersecurity field looks like. It also gives a brief description of the history and evolution of digital forensics as a field. Chapter 1 explains why the cybersecurity skills gap is both a challenge and an opportunity for students entering the field. Finally, we made a brief connection between cybersecurity, criminal justice, and digital forensics. Chapter 2 goes into detail about the connection between criminal justice and cybersecurity. Chapter 2 also points to how the skills gap challenge is an opportunity for students in the criminal justice field and why now is the time for criminal justice students to take advantage of this opportunity and make that career move.

Chapter 3 looks at an example of digital forensics as a career choice for criminal justice students. The chapter focuses on the various areas of digital forensics that a student in criminal justice can get involved in, from mobile devices to cloud forensics. Chapter 3 also discusses the valuable skills a digital forensics professional should have as well as a description of the digital forensics swim lanes (description of each type of digital forensic and the sources for digital evidence and information). Chapter 4 introduces the education and training that are available to students and what is needed in making a career decision. Chapter 5 identifies the opportunities available for criminal justice students within the field of cybersecurity and digital forensics. Chapter 5 also examines jobs within law enforcement and other sectors of the criminal justice system. Chapter 5 also covers available opportunities for students within law enforcement, the private sector, and the federal government.

Chapter 6 proposes a model or plan a criminal justice student can use to enter the cybersecurity/digital forensic field. This model/plan includes how a criminal justice student can enter the field within the components of the criminal justice system or as an alternative career option. The plan also includes details of what students should know as cybersecurity experts that will help them get hired. Chapter 7 provides students with the tools needed to prepare to search for, pursue, and attain employment and advancement in the cybersecurity and digital forensics field. Now that they have been presented with different career choices, how can they narrow down the job search? How can they get a foot in the door? What are strategies and tools they can use to develop their employment search in a more efficient and productive way, making the process smoother and getting the job that matters and utilizes the skill set they've just worked so hard to obtain?

This chapter concludes the book by summarizing the guidance provided in the book and listing questions students should ask themselves before they begin a career in digital forensics or cybersecurity. It also details career advancement suggestions and the benefits of a criminal justice perspective in the cybersecurity field.

Knowing all this information is crucial for those trying to decide whether to make this a career choice. However, there are other pertinent issues that a student has to consider as they make these decisions. They need to ask themselves certain questions, plan a career, and start thinking about retirement as well.

Questions Students Should Ask Themselves before They Begin a Career/as They Progress through Their Career

It is important for a student to ask themselves certain questions before embarking upon what we believe is an exciting career:

1. Do I see myself long-term doing this? Can I turn this opportunity into a career?
2. How can I contribute to the field? Where do I fit in? Can I make a strong connection between criminal justice and digital forensics or cybersecurity?
3. Do I have the skills to succeed in this career field? If not, what is my plan to acquire these skills?
4. Am I more suited to the public or private sector?
5. What will my five-, ten-, fifteen-, twenty-, or twenty-five-year career look like? What plans will I make toward a long career in the field?
6. When I retire, what will I do? Will I continue to advance in the field or teach/mentor the next generation?

Answering these questions will help students in determining how to navigate their career choices.

Tips for Advancement in the Cybersecurity and Digital Forensics Field

It is pertinent to note some very important tips that would help students navigate this interesting and diverse career field. These tips are drawn from previous chapters and other resources from career advisors. In chapter 3, we laid out the skills needed to advance your career in digital forensics. Chapters 4, 5, 6, and 7 guide students in the career decision with respect to what education and/or training as well as tools to use to assess information on getting a job. Students should revisit those chapters for specific information about education, training, certifications, and getting a job. Here we want to highlight or reemphasize certain tips.

Not all students are built for careers in a nine-to-five job. While in previous chapters, our emphasis was on getting jobs, there is also a place for novelty and entrepreneurship. The cybersecurity or digital forensics consultant field is a great option for students in criminal justice to innovate. Terhune and Hays, in their 2013 book *Land Your Dream Career in College*, state that it is important to innovate and to stand out. By this they mean college students should aspire to start a business where they can showcase the talents they have.[1] This is one way to separate yourself from all the other students out there who have the same skills and education you have. While there are obvious risks in starting a business that should not be ignored, there are also rewards, like fulfilling your dream, helping people or organizations fulfill their dreams, and working for yourself, rewards that can help you take that big step.

Terhune and Hays also emphasize networking. They describe networking simply as "to increase the amount of people in your circle or circles."[2] They reiterate that with each added person to your circle, you expand your circle and increase career opportunities tenfold. They also make a connection between networking and mentorship, as your mentor's circle could become part of your network.[3]

Along with networking comes mentorship. Identify a mentor or mentors who will help guide you through difficult career choices. Most mentors have experience with these choices, and while the generation gap may or may not be there, it is important to listen to your mentors. Developing a relationship with a mentor can be tricky, but as Terhune and Hays put it, "Everyone needs a 'soft place to fall,'"[4] meaning everyone needs someone to talk to, who will not judge them as they make difficult career decisions. Choose a mentor you can trust to be a cheerleader and a resourceful guide and who has traveled the way before you.

There are numerous tips you should keep in mind while you are still in school and as you journey through your career. Indeed.com gives fourteen great tips in advising students:[5]

1. Seek internship opportunities.
2. Consider taking part in a work-study program.
3. Grow your skills and knowledge.
4. Get an early start.
5. Keep your skills up-to-date.
6. Stay focused.
7. Find a balance with your personal life.
8. Pursue your passion.
9. Strive for excellence and stay motivated.
10. Use your school's career services.
11. Build your network.
12. Actively seek opportunities.
13. Create opportunities.
14. Find companies on social media.

These tips work for all careers; however, for the cybersecurity and digital forensic field, we would strongly emphasize networking, attending conferences where recruitment is taking place, and seeking internships and apprenticeships.

After a Cybersecurity Career, What Next?

A successful career hopefully sees you in a top or at least middle-management position in a reputable organization or managing your own company. But most of all, it is important for students to make career plans in increments of five years. While they may not necessarily know exactly what their career trajectory may be, it is important to have plans in place. Failing to plan is not an option. Planning helps professionals in answering the six questions in the previous section. Planning will also help in deciding things like what career path to follow, what skills need to be updated, when to update their skills, what training or certification they would need and when, at what level they see themselves in management roles, as well as when to retire.

Retirement: Was It All Worth It?

According to Sheridan and Rainville, everyone has to plan for retirement. They rightly assert that you will want to have a reasonable amount of income in retirement. They relate stories of how many people get to retirement and are shocked to find out that the amount of income they had established for retirement through pensions and social security is not enough to meet their needs. Therefore, they emphasize that every student needs a plan toward retirement from the day they start working.[6]

When you are in your twenties, retirement is the furthest thing from your mind. I know this was true for us, but planning now, while young, makes the transition that much easier. The cybersecurity/digital forensics world offers the ability, even when at retirement age, to continue to work and be productive. The skills of cybersecurity and digital forensics do not get old.

Complete List of Feeder Roles, According to Cyberseek.org

Job	Feeder Role(s)	Common Job Titles	Top Certifications	Top Skills
Cybersecurity/Specialist Technician (Entry Level)	Networking Systems Engineering Financial and Risk Analysis Security Intelligence	Information Security Specialist IT Security Specialist IT Specialist, Information Security Information Technology Specialist, Information Security	Certified Information Systems Security Professional (CISSP) SANS/GIAC Certification CompTIA Security+ Certified Information Systems Auditor (CISA) Certified Information Security Manager (CISM)	Information Security Information Systems Information Assurance Network Security Security Operations Vulnerability Assessment Project Management Linux NIST Cybersecurity Framework
Cybercrime Analyst/Cybercrime Investigator (Entry Level)	Networking Systems Engineering Financial and Risk Analysis Security Intelligence	Digital Forensics Analyst Cyber Forensic Specialist Cybersecurity Forensic Analyst Computer Forensics Analyst	SANS/GIAC Certification Certified Information Systems Security Professional (CISSP) EnCase Certified Examiner (EnCE) GIAC Certified Forensic Analyst GIAC Certified Incident Handler (GCIH)	Computer Forensics Linux Information Security Consumer Electronics Hard Drives Information Systems Forensic Toolkit UNIX Malware Engineering
Incident Analyst/Responder (Entry Level)	Networking Systems Engineering Financial and Risk Analysis Security Intelligence	Senior Analyst, Information Security Disaster Recovery Specialist Network Technical Specialist Audit Project Manager, Information Security	Certified Information Systems Security Professional (CISSP) SANS/GIAC Certification GIAC Certified Incident Handler (GCIH) CompTIA Security+ IT Infrastructure Library (ITIL) Certification	Information Security Project Management Information Systems Linux Network Security Technical Support Intrusion Detection UNIX Security Operations
IT Auditor (Entry Level)	Networking Systems Engineering	Senior IT Auditor IT Audit Consultant	Certified Information Systems Auditor (CISA)	Internal Auditing Audit Planning

Role	Field	Job Titles	Certifications	Knowledge / Skills
	Financial and Risk Analysis	IT Audit Manager IT Internal Auditor	Certified Information Systems Security Professional (CISSP) Information Systems Certification Certified Information Security Manager (CISM) IT Infrastructure Library (ITIL) Certification	Information Systems Sarbanes-Oxley (SOX) Accounting Risk Assessment Information Security COBIT Business Process
Cybersecurity Analyst (Mid-level)	Networking Software Development Systems Engineering	Information Security Analyst IT Security Analyst Cybersecurity Analyst Senior Security Analyst	Certified Information Systems Security Professional (CISSP) SANS/GIAC Certification Certified Information Systems Auditor (CISA) Certified Information Security Manager (CISM) CompTIA Security+	Information Security Information Systems Linux Network Security Threat Analysis Security Operations Vulnerability Assessment Project Management Intrusion Detection
Cybersecurity Consultant (Mid-level)		Security Specialist Security Consultant Physical Security Specialist Personnel Security Specialist	Certified Information Systems Security Professional (CISSP) Certified Information Systems Auditor (CISA) Certified Information Security Manager (CISM) SANS/GIAC Certification Information Systems Certification	Information Security Surveillance Information Systems Oracle Project Management Asset Protection Python Prevention of Criminal Activity Security Operations
Penetration and Vulnerability Tester (Mid-level)	Networking Software Development	Penetration Tester Senior Penetration Tester	Certified Information Systems Security Professional (CISSP)	Information Security Penetration Testing

Job	Feeder Role(s)	Common Job Titles	Top Certifications	Top Skills
Cybersecurity Manager/Administrator (Advanced Level)	Systems Engineering Software Development	Network Relations Consultant Application Security Analyst	SANS/GIAC Certification Certified Information Security Manager (CISM) Certified Information Systems Auditor (CISA) CompTIA Security+	Linux Python Java Vulnerability Assessment Information Systems Software Development Project Management
Cybersecurity Engineer (Advanced Level)	Networking Software Development Systems Engineering	Security Engineer Network Security Engineer Information Security Engineer Cyber Security Engineer	Certified Information Systems Security Professional (CISSP) SANS/GIAC Certification Certified Information Security Manager (CISM) CompTIA Security+ Certified Information Systems Auditor (CISA)	Information Security Network Security Linux Information Systems Python Cryptography Project Management Cisco Authentication
Cybersecurity Architect (Advanced Level)	Networking Software Development Systems Engineering	Security Architect IT Security Architect Senior Security Architect Cybersecurity Architect	Certified Information Systems Security Professional (CISSP) Certified Information Security Manager (CISM) SANS/GIAC Certification Certified Information Systems Auditor (CISA) IT Infrastructure Library (ITIL) Certification	Information Security Network Security Cryptography Information Systems Authentication Linux Software Development Cisco NIST Cybersecurity Framework

Note: Adapted from Cyberseek.org. Available at https://www.cyberseek.org/pathway.html.

Cybersecurity Roles Suitable for Criminal Justice Students, Adapted from Cyberseek.org

Feeder Roles	Entry-Level/ Mid-level Jobs	Common Job Titles	Certifications	Top Skills
Networking Systems Engineering Financial and Risk Analysis Security Intelligence	Cybercrime Analyst/ Investigator (Entry)	Digital Forensics Analyst Cyber Forensic Specialist Cybersecurity Forensic Analyst Computer Forensics Analyst	SANS/GIAC Certification Certified Information Systems Security Professional (CISSP) EnCase Certified Examiner (EnCE) GIAC Certified Forensic Analyst GIAC Certified Incident Handler (GCIH)	Digital Forensics Linux Information Security Consumer Electronics Hard Drives Information Systems Forensic Toolkit UNIX Malware Engineering
	Incident Analyst/ Responder (Entry)	Senior Analyst, Information Security Disaster Recovery Specialist Network Technical Specialist Audit Project Manager, Information Security	Certified Information Systems Security Professional (CISSP) SANS/GIAC Certification GIAC Certified Incident Handler (GCIH) CompTIA Security+ IT Infrastructure Library (ITIL) Certification	Information Security Project Management Information Systems Linux Network Security Technical Support Intrusion Detection UNIX Security Operations
Cybersecurity Consultant		Security Specialist Security Consultant Physical Security Specialist Personnel Security Specialist	Certified Information Systems Security Professional (CISSP) Certified Information Systems Auditor (CISA) Certified Information Security Manager (CISM) SANS/GIAC Certification Information Systems Certification	Security Operations Information Security Surveillance Information Systems Oracle Project Management Asset Protection Python Prevention of Criminal Activity Security Operations

Note: Adapted from Cyberseek.org. Available at https://www.cyberseek.org/pathway.html. Full table found in appendix 1: Cyberseek Roles.

Cybersecurity Roles for Criminal Justice Students, Adapted from the NIST SP 800-181

Investigate (IN)

NICE Specialty Area	NICE Specialty Area Description	Work Role	Work Role Description
Cyber Investigation (INV)	Applies tactics, techniques, and procedures for a full range of investigative tools and processes to include, but not be limited to, interview and interrogation techniques, surveillance, countersurveillance, and surveillance detection and appropriately balances the benefits of prosecution versus intelligence gathering.	Cybercrime Investigator	Identifies, collects, examines, and preserves evidence using controlled and documented analytical and investigative techniques.
Digital Forensics (FOR)	Collects, processes, preserves, analyzes, and presents computer-related evidence in support of network vulnerability mitigation and/or criminal, fraud, counterintelligence, or law enforcement investigations.	Law Enforcement/ Counterintelligence Forensics Analyst	Conducts detailed investigations on computer-based crimes establishing documentary or physical evidence, to include digital media and logs associated with cyber intrusion incidents.
		Cyber Defense Forensics Analyst	Analyzes digital evidence and investigates computer security incidents to derive useful information in support of system/ network vulnerability mitigation.

Protect and Defend (PR)

NICE Specialty Area	NICE Specialty Area Description	Work Role	Work Role Description
Cybersecurity Defense Analysis (CDA)	Uses defensive measures and information collected from a variety of sources to identify, analyze, and report events that occur or might occur within the network to protect information, information systems, and networks from threats.	Cyber Defense Analyst	Uses data collected from a variety of cyber defense tools (e.g., IDS alerts, firewalls, network traffic logs) to analyze events that occur within their environments for the purposes of mitigating threats.

Cybersecurity Defense Infrastructure Support (INF)	Tests, implements, deploys, maintains, reviews, and administers the infrastructure hardware and software that are required to effectively manage the computer network defense, service provider network, and resources. Monitors network to actively remediate unauthorized activities.	Cyber Defense Infrastructure Support Specialist	Tests, implements, deploys, maintains, and administers the infrastructure hardware and software.
Critical Incident Response (CIR)	Responds to crises or urgent situations within the pertinent domain to mitigate immediate and potential threats. Uses mitigation, preparedness, and response and recovery approaches, as needed, to maximize survival of life, preservation of property, and information security. Investigates and analyzes all relevant response activities.	Cyber Defense Incident Responder	Investigates, analyzes, and responds to cyber incidents within the network environment or enclave.
Vulnerability Assessment and Management (VAM)	Conducts assessments of threats and vulnerabilities; determines deviations from acceptable configurations, enterprise, or local policy; assesses the level of risk; and develops and/or recommends appropriate mitigation countermeasures in operational and nonoperational situations.	Vulnerability Assessment Analyst	Performs assessments of systems and networks within the network environment or enclave and identifies where those systems/networks deviate from acceptable configurations, enclave policy, or local policy. Measures effectiveness of defense-in-depth architecture against known vulnerabilities.

Collect and Operate (CO)

NICE Specialty Area	NICE Specialty Area Description	Work Role	Work Role Description
Collection Operations (CLO)	Executes collection using appropriate strategies and within the priorities established through the collection management process.	All Source-Collection Manager	Identifies collection authorities and environment; incorporates priority information requirements into collection management; develops concepts to meet leadership's intent. Determines capabilities of available collection assets; identifies new collection capabilities; and constructs and disseminates collection plans. Monitors execution of tasked collection to ensure effective execution of the collection plan.
		All Source-Collection Requirements Manager	Evaluates collection operations and develops effects-based collection requirements strategies using available sources and methods to improve collection. Develops, processes, validates, and coordinates submission of collection requirements. Evaluates performance of collection assets and collection operations.
Cyber Operational Planning (OPL)	Performs in-depth joint targeting and cybersecurity planning process. Gathers information and develops detailed operational plans and orders supporting requirements. Conducts strategic and operational-level planning across the full range of operations for integrated information and cyberspace operations.	Cyber Intel Planner	Develops detailed intelligence plans to satisfy cyber operations requirements. Collaborates with cyber operations planners to identify, validate, and levy requirements for collection and analysis. Participates in targeting selection, validation, synchronization, and execution of cyber actions. Synchronizes intelligence activities to support organization objectives in cyberspace.

	Cyber Ops Planner	Develops detailed plans for the conduct or support of the applicable range of cyber operations through collaboration with other planners, operators, and/or analysts. Participates in targeting selection, validation, and synchronization and enables integration during the execution of cyber actions.
	Partner Integration Planner	Works to advance cooperation across organizational or national borders between cyber operations partners. Aids the integration of partner cyber teams by providing guidance, resources, and collaboration to develop best practices and facilitate organizational support for achieving objectives in integrated cyber actions.
	Cyber Operator	Conducts collection, processing, and/or geolocation of systems to exploit, locate, and/or track targets of interest. Performs network navigation, tactical forensic analysis, and, when directed, executes on-net operations.
Cyber Operations (OPS)		Performs activities to gather evidence on criminal or foreign intelligence entities to mitigate possible or real-time threats; to protect against espionage or insider threats, foreign sabotage, or international terrorist activities; or to support other intelligence activities.

Analyze (AN)

NICE Specialty Area	NICE Specialty Area Description	Work Role	Work Role Description
Threat Analysis (TWA)	Identifies and assesses the capabilities and activities of cybersecurity criminals or foreign intelligence entities; produces findings to help initialize or support law enforcement and counterintelligence investigations or activities.	Threat/Warning Analyst	Develops cyber indicators to maintain awareness of the status of the highly dynamic operating environment. Collects, processes, analyzes, and disseminates cyber threat/warning assessments.
Exploitation Analysis (EXP)	Analyzes collected information to identify vulnerabilities and potential for exploitation.	Exploitation Analyst	Collaborates to identify access and collection gaps that can be satisfied through cyber collection and/or preparation activities. Leverages all authorized resources and analytic techniques to penetrate targeted networks.
All-Source Analysis (ASA)	Analyzes threat information from multiple sources, disciplines, and agencies across the intelligence community. Synthesizes and places intelligence information in context; draws insights about the possible implications.	All-Source Analyst	Analyzes data/information from one or multiple sources to conduct preparation of the environment, respond to requests for information, and submit intelligence collection and production requirements in support of planning and operations.
		Mission Assessment Specialist	Develops assessment plans and measures of performance/effectiveness. Conducts strategic and operational effectiveness assessments as required for cyber events. Determines whether systems performed as expected and provides input to the determination of operational effectiveness.

Category	Role	Description	
Targets (TGT)	Target Developer	Performs target system analysis and builds and/or maintains electronic target folders to include inputs from environment preparation and/or internal or external intelligence sources. Coordinates with partner target activities and intelligence organizations and presents candidate targets for vetting and validation.	Applies current knowledge of one or more regions, countries, nonstate entities, and/or technologies.
	Target Network Analyst	Conducts advanced analysis of collection and open-source data to ensure target continuity, to profile targets and their activities, and to develop techniques to gain more target information. Determines how targets communicate, move, operate, and live based on knowledge of target technologies, digital networks, and the applications on them.	
Language Analysis (LNG)	Multidisciplined Language Analyst	Applies language and culture expertise with target/threat and technical knowledge to process, analyze, and/or disseminate intelligence information derived from language, voice, and/or graphic material. Creates and maintains language-specific databases and working aids to support cyber action execution and ensure critical knowledge sharing. Provides subject matter expertise in foreign language-intensive or interdisciplinary projects.	Applies language, cultural, and technical expertise to support information collection, analysis, and other cybersecurity activities.

Oversee and Govern (OG)

NICE Specialty Area	NICE Specialty Area Description	Work Role	Work Role Description
Legal Advice and Advocacy (LGA)	Provides legally sound advice and recommendations to leadership and staff on a variety of relevant topics within the pertinent subject domain. Advocates legal and policy changes and makes a case on behalf of client via a wide range of written and oral work products, including legal briefs and proceedings.	Cyber Legal Advisor	Provides legal advice and recommendations on relevant topics related to cyber law.
		Privacy Officer/ Privacy Compliance Manager	Develops and oversees privacy compliance program and privacy program staff, supporting privacy compliance, governance/policy, and incident response needs of privacy and security executives and their teams.
Training, Education, and Awareness (TEA)	Conducts training of personnel within pertinent subject domain. Develops, plans, coordinates, delivers, and/or evaluates training courses, methods, and techniques as appropriate.	Cyber Instructional Curriculum Developer	Develops, plans, coordinates, and evaluates cyber training/education courses, methods, and techniques based on instructional needs.
		Cyber Instructor	Develops and conducts training or education of personnel within cyber domain.
Cybersecurity Management (MGT)	Oversees the cybersecurity program of an information system or network, including managing information security implications within the organization, specific program, or other area of responsibility, to include strategic, personnel, infrastructure, requirements, policy enforcement, emergency planning, security awareness, and other resources.	Information Systems Security Manager	Responsible for the cybersecurity of a program, organization, system, or enclave.
		Communications Security (COMSEC) Manager	Individual who manages the Communications Security (COMSEC) resources of an organization (CNSSI 4009) or key custodian for a Crypto Key Management System (CKMS).
Strategic Planning and Policy (SPP)	Develops policies and plans and/or advocates for changes in policy that support organizational cyberspace	Cyber Workforce Developer and Manager	Develops cyberspace workforce plans, strategies, and guidance to support cyberspace workforce manpower,

Specialty Area	Specialty Area Description	Work Role	Work Role Description
			personnel, and training and education requirements and to address changes to cyberspace policy, doctrine, materiel, force structure, and education and training requirements.
		Cyber Policy and Strategy Planner	Develops and maintains cybersecurity plans, strategy, and policy to support and align with organizational cybersecurity initiatives and regulatory compliance.
Executive Cyber Leadership (EXL)	Supervises, manages, and/or leads work and workers performing cyber and cyber-related and/or cyber operations work.	Executive Cyber Leadership	Executes decision-making authorities and establishes vision and direction for an organization's cyber and cyber-related resources and/or operations.
Program/Project Management (PMA) and Acquisition	Applies knowledge of data, information, processes, organizational interactions, skills, and analytical expertise, as well as systems, networks, and information exchange capabilities to manage acquisition programs. Executes duties governing hardware, software, and information system acquisition programs and other program management policies. Provides direct support for acquisitions that use information technology (IT) (including National Security Systems), applying IT-related laws and policies, and provides IT-related guidance throughout the total acquisition life cycle.	Program Manager	Leads, coordinates, communicates, integrates, and is accountable for the overall success of the program, ensuring alignment with agency or enterprise priorities.
		IT Project Manager	Directly manages information technology projects.
		Product Support Manager	Manages the package of support functions required to field and maintain the readiness and operational capability of systems and components.
		IT Investment/Portfolio Manager	Manages a portfolio of IT investments that align with the overall needs of mission and enterprise priorities.
		IT Program Auditor	Conducts evaluations of an IT program or its individual components to determine compliance with published standards.

Note: Adapted from the NIST SP 800-181—National Institute of Standards and Technology (NIST) Workforce Framework. Available at https://nvlpubs.nist.gov/nistpubs/SpecialPublications/NIST.SP.800-181.pdf. Complete list found on NICE Framework Specialty Areas and Work Role Table of Contents—Reference Spreadsheet.

Notes

CHAPTER 1. WHAT IS CYBERSECURITY?

1. National Initiative for Cybersecurity Careers and Studies, "Explore Terms: A Glossary of Common Cybersecurity Terminology," accessed October 20, 2019, https://niccs.us-cert.gov/about-niccs/cybersecurity-glossary.

2. Kaspersky, "What Is Cyber Security?" 2020, https://usa.kaspersky.com/resource-center/definitions/what-is-cyber-security.

3. Palo Alto Networks, "What Is Cybersecurity?" 2020, https://www.paloaltonetworks.com/cyberpedia/what-is-cyber-security.

4. Alice Grace Johansen, "What Is Cyber Security? What You Need to Know," Norton, July 24, 2020, https://us.norton.com/internetsecurity-malware-what-is-cybersecurity-what-you-need-to-know.html.

5. Janine Kremling and Amanda Parker Sharp, *Cyberspace, Cybersecurity, and Cybercrime*, first edition (Thousand Oaks, CA: Sage, 2018).

6. Response here does not mean incident response (IR). Although law enforcement agencies are tasked with investigating cybercrime, they are generally not first responders in the sense of IR. Law enforcement is usually alerted well after the crime has been committed.

7. David Wall, "Cybercrimes and Criminal Justice," *Criminal Justice Matters* 46, no. 1 (2001): 36–37.

8. *U.S. v. DePew*, 751 F. Supp. 1195, Casetext, accessed September 3, 2019, https://casetext.com/case/us-v-depew-6.

9. World Wide Web Foundation, "History of the Web," accessed September 3, 2019, https://webfoundation.org/about/vision/history-of-the-web/.

10. Michael A. Fuoco, "Missing Teen Found Safe but Tied Up in Virginia Townhouse," *Pittsburgh Post-Gazette*, January 5, 2002, http://old.post-gazette.com/regionstate/20020105missingp1.asp.

11. Apple, "Apple to Launch iCloud on October 12," October 4, 2011, https://www.apple.com/newsroom/2011/10/04Apple-to-Launch-iCloud-on-October-12/.

12. Federal Bureau of Investigation (FBI), "The Morris Worm," November 2, 2018, https://www.fbi.gov/news/stories/morris-worm-30-years-since-first-major-attack-on-internet-110218.

CHAPTER 2. THE CYBERSECURITY SKILLS GAP: AN OPPORTUNITY FOR CRIMINAL JUSTICE STUDENTS

1. Karen Evans and Franklin Reeder, "A Human Capital Crisis in Cybersecurity: Technical Proficiency Matters," Center for Strategic and International Studies, accessed November 6, 2019, https://www.csis.org/analysis/human-capital-crisis-cybersecurity.

2. Samantha A. Schwartz, "0% Unemployment Rate and 5 Other Numbers You Need to Know about Cybersecurity," CIODive, November 7, 2019, https://www.ciodive.com/news/0-unemployment-rate-and-5-other-numbers-you-need-to-know-about-cybersecuri/566779/.

3. Joe Scherrer, "Cybersecurity Engineering: A New Academic Discipline," VentureBeat, April 15, 2018, https://venturebeat.com/2018/04/15/cybersecurity-engineering-a-new-academic-discipline/; Johanna Trovato, "All Grown Up: Cybersecurity Moves into the Mainstream," Encoura, January 22, 2019, https://encoura.org/all-grown-up-cybersecurity-moves-into-the-mainstream/. Department of Homeland Security, "Cybersecurity Jobs," accessed November 6, 2019, https://www.dhs.gov/cisa/cybersecurity-jobs.

4. U.S. Bureau of Labor Statistics, "Fastest-Growing Occupations: Occupational Outlook Handbook," accessed November 6, 2019, https://www.bls.gov/ooh/fastest-growing.htm.

5. Evans and Reeder, "A Human Capital Crisis."

6. National Information Assurance Education and Training Program, "About CAE Program," accessed November 6, 2019, https://www.iad.gov/nietp/CAE Requirements.cfm.

7. Ibid.

8. Dan Lohrmann, "2019: The Year Ransomware Targeted State and Local Governments," *Government Technology*, December 23, 2019, https://www.govtech.com/blogs/lohrmann-on-cybersecurity/2019-the-year-ransomware-targeted-state—local-governments.html.

9. Bobby Allen, "22 Texas Towns Hit with Ransomware Attack in 'New Front' of Cyberassault," NPR, August 20, 2019, https://www.npr.org/2019/08/20/752695554/23-texas-towns-hit-with-ransomware-attack-in-new-front-of-cyberassault.

10. Niraj Chokshi, "Hackers Are Holding Baltimore Hostage: How They Struck and What's Next," *New York Times*, May 22, 2019, https://www.nytimes.com/2019/05/22/us/baltimore-ransomware.html.

11. Patricia Mazzei, "Hit by Ransomware Attack, Florida City Agrees to Pay Hackers $600,000," *New York Times*, June 19, 2019, https://www.nytimes.com/2019/06/19/us/florida-riviera-beach-hacking-ransom.html.

12. Margaret Baker, "Mississippi City Operations Disrupted by Ransomware Attack," Government Technology, December 10, 2019, https://www.govtech.com/security/Mississippi-City-Operations-Disrupted-by-Ransomware-Attack.html.

13. (ISC)², "Cybersecurity Professionals Focus on Developing New Skills as Workforce Gap Widens," 2018, https://www.isc2.org/-/media/ISC2/Research/2018-ISC2-Cybersecurity-Workforce-Study.ashx?la=en&hash=4E09681D0FB51698D9BA6BF13EEABFA48BD17DB0.

14. Aunshul Rege, "Leveraging Simulators to Educate Non-STEM Students in Conducting Real-time Cybersecurity Field Research," presented at the 2017 National Initiative for Cybersecurity Education (NICE) Conference, Dayton, Ohio, November 2017, https://www.nist.gov/system/files/documents/2020/08/14/1-1_Rege_PUBLIC.pdf.

15. U.S. Office of Personnel Management, "CyberCorps: Scholarship for Service," accessed November 9, 2020, https://www.sfs.opm.gov.

16. Tori Terhune and Betsy Hays, *Land Your Dream Career in College: A Complete Guide to Success* (Lanham, MD: Rowman & Littlefield., 2013).

CHAPTER 3. IT'S ALL ABOUT SKILLS

1. These tools are listed in alphabetical order. I am tool agnostic although I do have personal favorites.

2. CRU, "Ditto and Ditto DX," accessed September 11, 2019, https://www.cru-inc.com/ditto/.

3. Logicube, home page, accessed September 11, 2019, https://www.logicube.com/?v=7516fd43adaa.

4. OpenText, "Tableau Hardware Catalog," accessed September 11, 2019, https://www.guidancesoftware.com/tableau/hardware?types=Duplicators&cmpid=nav_r.

5. SolarWinds, "What Is NetFlow?" accessed September 12, 2019, https://www.solarwinds.com/netflow-traffic-analyzer/use-cases/what-is-netflow.

6. "RFC 7011—Specification of the IP Flow Information Export (IPFIX) Protocol for the Exchange of Flow Information," accessed September 12, 2019, https://tools.ietf.org/html/rfc7011.

7. Jonathan Chadwick, "Internet Encryption Hits 50%: Netflix Eating 15% of Global Traffic," *Tech Monitor*, accessed September 12, 2019, https://www .cbronline.com/news/internet-encryption-sandvine.

8. U.S. Department of Justice, "Promoting Public Safety, Privacy, and the Rule of Law Around the World: The Purpose and Impact of the CLOUD Act," April 2019, https://www.justice.gov/opa/press-release/file/1153446/download.

9. Canalys Newsroom, "Battle for Enterprise Cloud Customers Intensifies as Spending Grows 42% in Q1 2019," accessed September 23, 2019, https://www .canalys.com/newsroom/canalys-battle-for-enterprise-cloud-customers-intensifies -as-spending-grows-42-in-q1-2019.

10. A virtual machine (VM) is a computer that exists entirely in software. It looks and acts just like a circuit-based computer, but it's all software. VMs require a controlling program, aka a hypervisor, that provides hardware resources to the VM. Some popular virtual environments include VMWare, Hyper-V, Xen, KVM, etc.

11. Aravind Swaminathan, Robert Loeb, and Emily S. Tabatabai, "The CLOUD Act, Explained," Orrick, April 6, 2018, https://www.orrick.com/%20 Insights/2018/04/The-CLOUD-Act-Explained.

12. TechJury, "Gmail Statistics and Trends: What You Need to Know in 2019," accessed September 24, 2019, https://techjury.net/stats-about/gmail -statistics/.

13. Ascii (American Standard Code for Information Interchange) represents human readable characters and each character is 8 bits or 1 byte. Unicode was developed by Microsoft and seeks to solve the challenge of multiple languages. Unicode characters are represented by 16 bits or 2 bytes.

14. LoveToKnow, "How Many Cell Phones Are in the U.S.?" accessed September 24, 2019, https://cellphones.lovetoknow.com/how-many-cell-phones-are-us.

15. We Are Social, "Digital 2019," accessed September 24, 2019, https:// wearesocial.com/blog/2019/01/digital-2019-global-internet-use-accelerates.

16. MacRumors, "Apple Now Has 1.4 Billion Active Devices Worldwide," accessed September 24, 2019, https://www.macrumors.com/2019/01/29/apple -1-4-billion-active-devices/.

17. Russell Brandom, "There Are Now 2.5 Billion Active Android Devices," *The Verge*, May 7, 2019, https://www.theverge.com/2019/5/7/18528297/google -io-2019-android-devices-play-store-total-number-statistic-keynote.

18. Given the fact that there are many Android phone vendors, it is not possible to determine if all encrypt by default. Google, with Android 5.0 Lollipop, says the encryption must be turned on by default. John Zorabedian, "New Android Marshmallow Devices Must Have Default Encryption, Google Says," Naked

Security, October 21, 2015, https://nakedsecurity.sophos.com/2015/10/21/new
-android-marshmallow-devices-must-have-default-encryption-google-says/.

19. GitHub, "Ipwndfu/Checkm8.Py at Master · Axi0mX/Ipwndfu · GitHub,"
accessed October 2, 2019, https://github.com/axi0mX/ipwndfu/blob/master/
checkm8.py.

20. Some incorporate decompilation in the category of static analysis. It is
separated here due to skill level requirements.

21. Base64 is the algorithm used to transmit binary files via email on the internet. Base64 converts the binary into characters; then when the file is received on
the other side, the file is converted back to its original format.

22. Statista, "Desktop OS Market Share 2013–2019," accessed October
14, 2019, https://www.statista.com/statistics/218089/global-market-share-of-win
dows-7/.

23. DOS stands for Disk Operating Systems and was Microsoft's first operating system and ran on IBM personal computers (PCs).

24. SecurityWeek, "Hackers Are Loving PowerShell, Study Finds," accessed
October 14, 2019, https://www.securityweek.com/hackers-are-loving-powershell
-study-finds.

25. .NET is a software development platform designed by Microsoft. If you
can't do it in .NET, you can't do it.

26. Codrut Neagu, "Simple Questions: What Is NTFS and Why Is It Useful?"
Digital Citizen, accessed October 14, 2019, https://www.digitalcitizen.life/what
-is-ntfs-why-useful.

27. Linux Training Academy, "What Is Linux?" accessed October 14, 2019,
https://www.linuxtrainingacademy.com/what-is-linux/.

28. https://www.kali.org/.

29. http://sumuri.com/product-category/brands/paladin/.

30. I have had some trouble with Windows large formatted ex-FAT drives being readable on Mac OS.

31. There are many great TCP/IP references, but if you wish to pursue TCP/
IP further, I recommend: Douglas E. Commer, *Internetworking with TCP/IP*, fifth
edition (Upper Saddle River, NJ: Pearson/Prentice Hall, 2006).

32. Switching can also be used to move data between networks with protocols
such as MPLS (multi-protocol label switching).

33. MPLS or Multi-Protocol Label Switching is a wide area or network-to-
network switching technology.

34. Cultural variations will change your approach.

35. A write-blocker is an electronic device that you attach to a piece of media
(hard drive, thumb drive, etc.) that prevents a computer from writing to the drive,
making the drive read only.

36. SHA stands for secure hashing algorithm. There are multiple versions of SHA: SHA 1, SHA 2 (224-bit, 256-bit, 384-bit, or 512 versions). There is even a SHA 3.

37. The going concern issue here is that the seizure of the hardware could put the company out of business.

CHAPTER 4. EDUCATION AND CERTIFICATIONS

1. A Venn diagram is a graphical representation, usually a few circles, that are used to show the relationship between different things or sets of data.

2. Do not confuse security with chain of custody. Digital forensics is directly linked to and dependent on chain of custody. Chain of custody is the process of guaranteeing the integrity of evidence from the moment it is first collected until its eventual disposition.

3. I purchased twelve Cisco 2811 routers for $80 each.

4. All of these courses do not necessarily need to be required courses but do need to be offered.

5. The website had this school listed as University of Maryland University College. The school changed its name on July 1, 2019.

6. The website had this school listed as Sam Houston University.

7. It is important to note here that just because a specific institution is not mentioned does not mean that they are not noteworthy. It is important to apply the list of components of a digital forensics program to any institution before making a decision.

CHAPTER 5. CYBERSECURITY CAREER OPPORTUNITIES IN THE FIELD OF CRIMINAL JUSTICE

1. Ronny Richardson and Max M. North, "Ransomware: Evolution, Mitigation and Prevention," *Faculty Publications* 13, no. 1 (January 1, 2017): 10–21.

2. Dan Lohrmann, "How Local Governments Can Address Cybersecurity Challenges," *Government Technology*, August 4, 2019, https://www.govtech.com/ blogs/lohrmann-on-cybersecurity/how-local-governments-can-address-cybersecu rity.html.

3. Thomas J. Holt, *Crime Online: Correlates, Causes and Context*, third edition (Durham, NC: Carolina Academic Press, 2016).

4. Janine Kremling and Amanda Parker Sharp, *Cyberspace, Cybersecurity, and Cybercrime*, first edition (Thousand Oaks, CA: Sage, 2018).

5. Police Executive Research Forum, "Critical Issues in Policing Series: The Role of Local Law Enforcement Agencies in Preventing and Investigating Cyber-

crime," April 30, 2014, https://www.policeforum.org/assets/docs/Critical_Issues_Series_2/the%20role%20of%20local%20law%20enforcement%20agencies%20in%20preventing%20and%20investigating%20cybercrime%202014.pdf.

6. Texas Department of Public Safety, "Computer Information Technology and Electronic Crime (CITEC) Unit," accessed November 9, 2020, https://www.dps.texas.gov/CriminalInvestigations/citecUnit.htm.

7. Ibid.

8. City of Houston, "Houston Information Technology Services," accessed November 9, 2020, https://www.houstontx.gov/hits/.

9. City of Austin, "Information Security," accessed November 9, 2020, https://data.austintexas.gov/stories/s/Information-Security/i2hy-f2ey/.

10. City of Dallas, "Stop. Think. Connect," accessed November 9, 2020, https://dallascityhall.com/departments/officeemergencymanagement/Pages/Stop-Think-Connect.aspx.

11. Houston Police Department, "Cyber and Financial Crimes," accessed November 9, 2020, https://www.houstontx.gov/police/divisions/cyber_&_financial_crimes/index.htm.

12. City of Houston, "Houston Information Technology Services."

13. Chuck Easttom, *System Forensics, Investigation, and Response*, third edition (Burlington, MA: Jones and Bartlett Learning, 2019).

14. Cyberseek, "Cybersecurity Career Pathway," accessed February 20, 2020, https://www.cyberseek.org/pathway.html.

15. Ibid.

16. William Newhouse et al., "National Initiative for Cybersecurity Education (NICE) Cybersecurity Workforce Framework," NIST Special Publication 800-181, 2017, https://doi.org/10.6028/NIST.SP.800-181.

CHAPTER 6. PLANNING YOUR PATH INTO THE CYBERSECURITY AND DIGITAL FORENSICS FIELD

1. Candy Alexander, "The Cybersecurity Skills Gap," *SC Magazine*, December 8, 2014, https://www.scmagazine.com/home/opinions/the-cybersecurity-skills-gap/.

2. Candy Alexander, "The Cybersecurity Profession Has A Clear Career Path. LOL. Just Kidding," *At the Intersection of Technology, Cybersecurity, and Society Magazine*, November 30, 2016, https://www.itspmagazine.com/from-the-newsroom/the-cybersecurity-profession-has-a-clear-career-path-lol-just-kidding.

3. Courtney Connley, "Google, Apple, and 12 Other Companies That No Longer Require Employees to Have a College Degree," *CNBC Careers*, October 8, 2018, https://www.cnbc.com/2018/08/16/15-companies-that-no-longer-require-employees-to-have-a-college-degree.html; Glassdoor Team, "15 More Companies

That No Longer Require a Degree—Apply Now," January 10, 2020. https://www.glassdoor.com/blog/no-degree-required/.

4. Jennifer Bonds-Raacke, John Raacke, and Samantha Elliott, "Should I Be Networking? Exploring the Importance of Networking for Students: Opportunity Knocks through Relationship Building," APA, January 2017. https://www.apa.org/ed/precollege/psn/2017/01/importance-networking.

5. Lexico, "Definition of Internship in English," accessed June 21, 2020, https://www.lexico.com/en/definition/internship.

6. Apprenticeship.gov, "Jump Start Your Career through Apprenticeship: What Is an Apprenticeship?" https://www.apprenticeship.gov/become-apprentice.

7. Department of Homeland Security (DHS), "Cybersecurity Competitions," accessed June 21, 2020, https://www.dhs.gov/science-and-technology/cybersecurity-competitions#.

8. Lucy Tsado, "Cybersecurity Education: The Need for a Top-Driven, Multidisciplinary, School-Wide Approach," *Journal of Cybersecurity Education, Research and Practice* 4, no. 1 (June 2019), https://digitalcommons.kennesaw.edu/jcerp/vol2019/iss1/4/.

CHAPTER 7. GETTING THE JOB AND ENTERING THE DIGITAL FORENSICS FIELD

1. In the world of digital forensics, the term *media* also refers to physical storage such as hard drives, thumb drives, SD cards, DVDs, etc. So when talking about media, you need to understand the context of how it's used.

2. Command line version of Autopsy.

3. A hash is a digital fingerprint that is unique to the file from which was derived. MD-5, SHA-1/256/512 are examples of hashing algorithms.

4. Unicode was developed by Microsoft and allows Microsoft Windows to handle multiple languages.

5. A signature is several artifacts that, when combined, point to something of interest (e.g., malware).

6. Android is one of the two major operating systems used in mobile devices. The other is Apple's IOS.

7. This is how you get the data into Autopsy or any digital forensics tools for that matter. Raw/dd originally was derived from the Linux dd program. The E01 format was originally developed by Guidance Software/OpenText for EnCase.

8. A jailbroken phone is one where its operating system has been compromised or rooted thereby rendering the security features of the phone ineffective.

9. U.S. Office of Personnel Management, "Direct Hire Authority," accessed February 4, 2020, https://www.opm.gov/policy-data-oversight/hiring-information/direct-hire-authority/.

10. HSI is an agency within the Department of Homeland Security (DHS) that specializes in investigating federal crimes to include export enforcement, immigration fraud, cybercrime, international crime, etc. (https://www.ice.gov/hsi).

11. USAJOBS Help Center, "What Is a Series or Grade?," accessed February 4, 2020, https://www.usajobs.gov/Help/faq/pay/series-and-grade/.

CONCLUSION: CAREER ADVANCEMENT IN CYBERSECURITY AND DIGITAL FORENSICS

1. Tori Terhune and Betsy Hays, *Land Your Dream Career in College: A Complete Guide to Success* (Lanham, MD: Rowman & Littlefield, 2013).

2. Tori Terhune and Betsy Hays, *Life after College: Ten Steps to Build the Life You Love* (Lanham, MD: Rowman & Littlefield, 2014), 131.

3. Terhune and Hays, *Land Your Dream Career.*

4. Terhune and Hays, *Life after College,* 168.

5. Indeed, "Indeed Career Guide: 14 Career Advice Tips for College Students," May 28, 2020, https://www.indeed.com/career-advice/career-development/career-advice-for-college-students.

6. Matthew Sheridan and Raymond Rainville, *Exploring Careers in Criminal Justice: A Comprehensive Guide* (Lanham, MD: Rowman & Littlefield, 2016).

Bibliography

Alexander, Candy. "The Cybersecurity Profession Has A Clear Career Path. LOL. Just Kidding." *At the Intersection of Technology, Cybersecurity, and Society Magazine.* November 30, 2016. https://www.itspmagazine.com/from-the-news room/the-cybersecurity-profession-has-a-clear-career-path-lol-just-kidding.

———. "The Cybersecurity Skills Gap." *SC Magazine.* December 8, 2014. https://www.scmagazine.com/home/opinions/the-cybersecurity-skills-gap/.

Allen, Bobby. "22 Texas Towns Hit with Ransomware Attack in 'New Front' of Cyberassault." NPR. August 20, 2019. https://www.npr.org/2019/08/20/752695554/23-texas-towns-hit-with-ransomware-attack-in-new-front-of-cyberassault.

Apple. "Apple to Launch iCloud on October 12." October 4, 2011. https://www.apple.com/newsroom/2011/10/04Apple-to-Launch-iCloud-on-October-12/.

Apprenticeship.gov. "Jump Start Your Career through Apprenticeship: What Is an Apprenticeship?" https://www.apprenticeship.gov/become-apprentice.

Baker, Margaret. "Mississippi City Operations Disrupted by Ransomware Attack." Government Technology, December 10, 2019. https://www.govtech.com/security/Mississippi-City-Operations-Disrupted-by-Ransomware-Attack.html.

Bonds-Raacke, Jennifer, John Raacke, and Samantha Elliott. "Should I Be Networking? Exploring the Importance of Networking for Students: Opportunity Knocks through Relationship Building." APA, January 2017. https://www.apa.org/ed/precollege/psn/2017/01/importance-networking.

Brandom, Russell. "There Are Now 2.5 Billion Active Android Devices." *The Verge.* May 7, 2019. https://www.theverge.com/2019/5/7/18528297/google-io-2019-android-devices-play-store-total-number-statistic-keynote.

Canalys Newsroom. "Battle for Enterprise Cloud Customers Intensifies as Spending Grows 42% in Q1 2019." Accessed September 23, 2019. https://www.canalys.com/newsroom/canalys-battle-for-enterprise-cloud-customers-intensifies-as-spending-grows-42-in-q1-2019.

Chadwick, Jonathan. "Internet Encryption Hits 50%: Netflix Eating 15% of Global Traffic." *Tech Monitor*. Accessed September 12, 2019. https://www.chronline.com/news/internet-encryption-sandvine.

Chokshi, Niraj. "Hackers Are Holding Baltimore Hostage: How They Struck and What's Next." *New York Times*. May 22, 2019. https://www.nytimes.com/2019/05/22/us/baltimore-ransomware.html.

City of Austin. "Information Security." Accessed November 9, 2020. https://data.austintexas.gov/stories/s/Information-Security/i2hy-f2ey/.

City of Dallas. "Stop. Think. Connect." Accessed November 9, 2020. https://dallascityhall.com/departments/officeemergencymanagement/Pages/Stop-Think-Connect.aspx.

City of Houston. "Houston Information Technology Services." Accessed November 9, 2020. https://www.houstontx.gov/hits/.

Connley, Courtney. "Google, Apple, and 12 Other Companies That No Longer Require Employees to Have a College Degree." *CNBC Careers*. October 8, 2018. https://www.cnbc.com/2018/08/16/15-companies-that-no-longer-require-employees-to-have-a-college-degree.html.

CRU. "Ditto and Ditto DX." Accessed September 11, 2019. https://www.cru-inc.com/ditto/.

Cyberseek. "Cybersecurity Career Pathway." Accessed February 20, 2020. https://www.cyberseek.org/pathway.html.

Department of Homeland Security (DHS). "Cybersecurity Competitions." Accessed June 21, 2020. https://www.dhs.gov/science-and-technology/cybersecurity-competitions#.

———. "Cybersecurity Jobs." Accessed November 6, 2019. https://www.dhs.gov/cisa/cybersecurity-jobs.

Easttom, Chuck. *System Forensics, Investigation, and Response*. Third edition. Burlington, MA: Jones and Bartlett Learning, 2019.

Evans, Karen, and Franklin Reeder. "A Human Capital Crisis in Cybersecurity: Technical Proficiency Matters." Center for Strategic and International Studies. Accessed November 6, 2019. https://www.csis.org/analysis/human-capital-crisis-cybersecurity.

Federal Bureau of Investigation (FBI). "The Morris Worm." November 2, 2018. https://www.fbi.gov/news/stories/morris-worm-30-years-since-first-major-attack-on-internet-110218.

Freidberg, Barbara. "Jobs With the Lowest Unemployment Rates." Investopedia. October 7, 2018. https://www.investopedia.com/personal-finance/jobs-lowest-unemployment-rates/.

Fuoco, Michael A. "Missing Teen Found Safe but Tied Up in Virginia Town-house." *Pittsburgh Post-Gazette*. January 5, 2002. http://old.post-gazette.com/regionstate/20020105missingp1.asp.

GitHub. "Ipwndfu/Checkm8.Py at Master · Axi0mX/Ipwndfu." Accessed October 2, 2019. https://github.com/axi0mX/ipwndfu/blob/master/checkm8.py.

Glassdoor Team. "15 More Companies That No Longer Require a Degree—Apply Now." January 10, 2020. https://www.glassdoor.com/blog/no-degree-required/.

Holt, Thomas J. *Crime Online: Correlates, Causes and Context*. Third edition. Durham, NC: Carolina Academic Press, 2016.

Houston Police Department. "Cyber and Financial Crimes." Accessed November 9, 2020. https://www.houstontx.gov/police/divisions/cyber_&_financial_crimes/index.htm.

Indeed. "Indeed Career Guide: 14 Career Advice Tips for College Students." May 28, 2020. https://www.indeed.com/career-advice/career-development/career-advice-for-college-students.

(ISC)². "Cybersecurity Professionals Focus on Developing New Skills as Work-force Gap Widens." 2018. https://www.isc2.org/-/media/ISC2/Research/2018-ISC2-Cybersecurity-Workforce-Study.ashx?la=en&hash=4E09681D0FB51698D9BA6BF13EEABFA48BD17DB0.

Johansen, Alice Grace. "What Is Cyber Security? What You Need to Know." Norton. July 24, 2020. https://us.norton.com/internetsecurity-malware-what-is-cybersecurity-what-you-need-to-know.html.

Kaspersky. "What Is Cyber Security?" 2020. https://usa.kaspersky.com/resource-center/definitions/what-is-cyber-security.

Kremling, Janine, and Amanda Parker Sharp. *Cyberspace, Cybersecurity, and Cyber-crime*. First edition. Thousand Oaks, CA: Sage, 2018.

Lexico. "Definition of Internship in English." Accessed June 21, 2020. https://www.lexico.com/en/definition/internship.

Linux Training Academy. "What Is Linux?" Accessed October 14, 2019. https://www.linuxtrainingacademy.com/what-is-linux/.

Logicube. Home page. Accessed September 11, 2019. https://www.logicube.com/?v=7516fd43adaa.

Lohrmann, Dan. "2019: The Year Ransomware Targeted State and Local Governments." *Government Technology*. December 23, 2019. https://www.govtech.com/blogs/lohrmann-on-cybersecurity/2019-the-year-ransomware-targeted-state—local-governments.html.

———. "How Local Governments Can Address Cybersecurity Challenges." *Government Technology*. August 4, 2019. https://www.govtech.com/blogs/lohrmann-on-cybersecurity/how-local-governments-can-address-cybersecurity.html.

LoveToKnow. "How Many Cell Phones Are in the U.S.?" Accessed September 24, 2019. https://cellphones.lovetoknow.com/how-many-cell-phones-are-us.

MacRumors. "Apple Now Has 1.4 Billion Active Devices Worldwide." Accessed September 24, 2019. https://www.macrumors.com/2019/01/29/apple-1-4-billion-active-devices/.

Mazzei, Patricia. "Hit by Ransomware Attack, Florida City Agrees to Pay Hackers $600,000." *New York Times*. June 19, 2019. https://www.nytimes.com/2019/06/19/us/florida-riviera-beach-hacking-ransom.html.

National Information Assurance Education and Training Program. "About CAE Program." Accessed November 6, 2019. https://www.iad.gov/nietp/CAE Requirements.cfm.

———. "NSA/DHS National CAE in Cyber Defense Designated Institutions." Accessed June 22, 2020. https://www.iad.gov/NIETP/documents/Requirements/CAE-CD_2020_Knowledge_Units.pdf.

National Initiative for Cybersecurity Careers and Studies. "Explore Terms: A Glossary of Common Cybersecurity Terminology." Accessed October 20, 2019. https://niccs.us-cert.gov/about-niccs/cybersecurity-glossary.

Neagu, Codrut. "Simple Questions: What Is NTFS and Why Is It Useful?" Digital Citizen. Accessed October 14, 2019. https://www.digitalcitizen.life/what-is-ntfs-why-useful.

Newhouse, William, Stephanie Keith, Benjamin Scribner, and Greg Witte. "National Initiative for Cybersecurity Education (NICE) Cybersecurity Workforce Framework." NIST Special Publication 800-181. 2017. https://doi.org/10.6028/NIST.SP.800-181.

OpenText. "Tableau Hardware Catalog." Accessed September 11, 2019. https://www.guidancesoftware.com/tableau/hardware?types=Duplicators&cmpid=nav_r.

Palo Alto Networks. "What Is Cybersecurity?" 2020. https://www.paloaltonetworks.com/cyberpedia/what-is-cyber-security.

Police Executive Research Forum. "Critical Issues in Policing Series: The Role of Local Law Enforcement Agencies in Preventing and Investigating Cybercrime." April 30, 2014. https://www.policeforum.org/assets/docs/Critical_Issues_Series_2/the%20role%20of%20local%20law%20enforcement%20agencies%20in%20preventing%20and%20investigating%20cybercrime%202014.pdf.

Rege, Aunshul. "Leveraging Simulators to Educate Non-STEM Students in Conducting Real-time Cybersecurity Field Research." Presented at the 2017 National Initiative for Cybersecurity Education (NICE) Conference, Dayton, Ohio, November 2017. https://www.nist.gov/system/files/documents/2020/08/14/1-1_Rege_PUBLIC.pdf.

"RFC 7011—Specification of the IP Flow Information Export (IPFIX) Protocol for the Exchange of Flow Information." Accessed September 12, 2019. https://tools.ietf.org/html/rfc7011.

Richardson, Ronny, and Max M. North. "Ransomware: Evolution, Mitigation and Prevention." *Faculty Publications* 13, no. 1 (January 1, 2017): 10–21.

Scherrer, Joe. "Cybersecurity Engineering: A New Academic Discipline." VentureBeat." April 15, 2018. https://venturebeat.com/2018/04/15/cybersecurity
-engineering-a-new-academic-discipline/.

Schwartz, Samantha A. "0% Unemployment Rate and 5 Other Numbers You Need to Know about Cybersecurity." CIODive. November 7, 2019. https://
www.ciodive.com/news/0-unemployment-rate-and-5-other-numbers-you
-need-to-know-about-cybersecuri/566779/.

SecurityWeek. "Hackers Are Loving PowerShell, Study Finds." Accessed October 14, 2019. https://www.securityweek.com/hackers-are-loving-powershell
-study-finds.

Sheridan, Matthew, and Raymond Rainville. *Exploring Careers in Criminal Justice: A Comprehensive Guide.* Lanham, MD: Rowman & Littlefield, 2016.

SolarWinds. "What Is NetFlow?" Accessed September 12, 2019. https://www
.solarwinds.com/netflow-traffic-analyzer/use-cases/what-is-netflow.

Spencer, Mark. "Beyond Timelines—Anchors in Relative Time." *Digital Forensics Magazine*, no. 18 (February 2014): 15–19.

Statista. "Desktop OS Market Share 2013–2019." Accessed October 14, 2019. https://www.statista.com/statistics/218089/global-market-share-of-windows-7/.

Swaminathan, Aravind, Robert Loeb, and Emily S. Tabatabai. "The CLOUD Act, Explained." Orrick. April 6, 2018. https://www.orrick.com/%20In
sights/2018/04/The-CLOUD-Act-Explained.

TechJury. "Gmail Statistics and Trends: What You Need to Know in 2019." Accessed September 24, 2019. https://techjury.net/stats-about/gmail-statistics/.

Terhune, Tori, and Betsy Hays. *Land Your Dream Career in College: A Complete Guide to Success.* Lanham, MD: Rowman & Littlefield, 2013.

———. *Life after College: Ten Steps to Build the Life You Love.* Lanham, MD: Rowman & Littlefield, 2014.

Texas Department of Public Safety. "Computer Information Technology and Electronic Crime (CITEC) Unit." Accessed November 9, 2020. https://www
.dps.texas.gov/CriminalInvestigations/citecUnit.htm.

Trovato, Johanna. "All Grown Up: Cybersecurity Moves into the Mainstream." Encoura. January 22, 2019. https://encoura.org/all-grown-up-cybersecurity
-moves-into-the-mainstream/.

Tsado, Lucy. "Cybersecurity Education: The Need for a Top-Driven, Multidisciplinary, School-Wide Approach." *Journal of Cybersecurity Education, Research and Practice* 4, no. 1 (June 2019). https://digitalcommons.kennesaw.edu/jcerp/
vol2019/iss1/4/.

U.S. Bureau of Labor Statistics. "Fastest-Growing Occupations: Occupational Outlook Handbook." Accessed November 6, 2019. https://www.bls.gov/ooh/
fastest-growing.htm.

U.S. Department of Justice. "Order Compelling Apple, Inc. to Assist Agents in Search." February 16, 2016. https://www.justice.gov/usao-cdca/file/825001/download.

———. "Promoting Public Safety, Privacy, and the Rule of Law Around the World: The Purpose and Impact of the CLOUD Act." April 2019. https://www.justice.gov/opa/press-release/file/1153446/download.

U.S. Office of Personnel Management. "CyberCorps: Scholarship for Service." Accessed November 9, 2020. https://www.sfs.opm.gov.

———. "Direct Hire Authority." Accessed February 4, 2020. https://www.opm.gov/policy-data-oversight/hiring-information/direct-hire-authority/.

U.S. v. DePew, 751 F. Supp. 1195. Casetext. Accessed September 3, 2019. https://casetext.com/case/us-v-depew-6.

USAJobs Help Center. "What Is a Series or Grade?" Accessed February 4, 2020. https://www.usajobs.gov/Help/faq/pay/series-and-grade/.

Wall, David. "Cybercrimes and Criminal Justice." *Criminal Justice Matters* 46, no. 1 (2001): 36–37.

We Are Social. "Digital 2019." Accessed September 24, 2019. https://wearesocial.com/blog/2019/01/digital-2019-global-internet-use-accelerates.

World Wide Web Foundation. "History of the Web." Accessed September 3, 2019. https://webfoundation.org/about/vision/history-of-the-web/.

Zorabedian, John. "New Android Marshmallow Devices Must Have Default Encryption, Google Says." Naked Security. October 21, 2015. https://nakedsecurity.sophos.com/2015/10/21/new-android-marshmallow-devices-must-have-default-encryption-google-says/.

Index

About the Authors

Lucy K. Tsado, PhD, is an assistant professor in the Department of Sociology, Social Work, and Criminal Justice at Lamar University, in Beaumont, Texas. Her interests include finding ways organizations, educational institutions, students, and local communities can work together to fill the cybersecurity skills gap. She believes in the ability of technical and nontechnical, as well as traditional and nontraditional, students to explore careers in cybersecurity and digital forensics to help fill that skills gap. At Lamar, Tsado teaches cybersecurity; digital forensics; cybercrime; corrections; criminal justice policy, planning, and evaluation; and class, race, gender, and crime to criminal justice students.

Robert Osgood (Bob), an engineer/CPA, is a twenty-six-year veteran FBI computer forensics examiner and Technically Trained Special Agent. His specialties include digital forensics, data intercept, cybercrime, enterprise criminal organizations, espionage, and counterterrorism. In the course of his work, he has performed digital forensics research and development and created unique new software tools for digital forensic law enforcement. He also serves as a digital forensics consultant to Probity, Inc., working with the Truxton development team. Osgood formed the first FBI computer forensics squad in 2000, served as the chief of the FBI's Digital Media Exploitation Unit, and was part of the team that executed the first court-authorized digital computer intercept at the FBI.

He managed and deployed the Washington, D.C., gunshot detection system. Osgood teaches digital media forensics, network forensics, incident response, digital forensics analysis, and fraud analytics, and he has taught digital forensics and cybercrime internationally. He is director of the MS in Digital Forensics program at George Mason University.